QUANTUM PHYSICS FOR BEGINNERS

DISCOVER THE SCIENCE OF QUANTUM MECHANICS AND LEARN THE MOST IMPORTANT CONCEPTS CONCERNING BLACK HOLES, STRINGS THEORY, AND WHAT WE PERCEIVE AS REALITY IN A SIMPLE WAY

Edwin Hines

Contents

Introduction	v
1. WHAT IS QUANTUM PHYSICS AND QUANTUM MECHANICS?	1
What is Quantum Physics?	0
How Does Quantum Physics Differ from Classical Physics?	0
Some Essential Characteristics of Quantum Physics	1
What is Quantum Mechanics?	2
2. THREE PRINCIPLES OF TRANSFORMATION	7
3. WHAT ARE PARTICLES OF LIGHT?	12
Understanding the Quantum Nature of Light	12
Light Waves	17
Interference	18
Light Quanta	18
4. WHAT IS THE WAVE OF MATTER?	21
5. PRINCIPLE OF UNCERTAINTY	24
6. DIFFERENCES BETWEEN QUANTUM PHYSICS AND GENERAL RELATIVITY	28
Quantum Mechanics and General Relativity Incompatibility	32
7. THE SCHRODINGER'S CAT	34
8. THE BUILDING BLOCKS OF MATTER AND WAVE-PARTICLE DUALITY	36
9. QUANTUM POSSIBILITIES AND WAVES	39
What is Wave Interference?	42
10. "DARK BODY" SPECTRUM UNDERSTANDING THE CURVE OF THE BLACK BODY	44
Heated Bodies Radiate	44
How Is Radiation Absorbed?	45

Absorption and Excretion	46
Two Basic Rules	47
11. ENTANGLEMENT	49
12. AN INTRODUCTION TO THE STRINGS THEORY MADE EASY FOR BEGINNERS	52
13. BLACK HOLES AND SPACETIME	55
Black Holes and the Mystery of Quantum Gravity	55
14. MOST POPULAR EXPERIMENTS EXPLAINED	59
Unified Energy and Unified Matter	59
Superconductors	61
15. QUANTUM COMPUTING	68
What is a Computer?	68
How do Computers Work?	69
What is a Qubit?	72
How Would Quantum Computers Operate?	76
16. WAYS QUANTUM PHYSICS AFFECTS YOUR DAILY LIFE	79
The Neon Light	79
The Laser	81
The GPS	83
The Anti-Gravity Wheel	86
The Semiconductor	87
The Solar Panel and the Light-Emitting Diode (LED)	89
Superconductivity	90
Quantum Physics as Seen in Everyday Objects	92
Afterword	95

Introduction

Quantum physics is an interesting physics branch that deals with the microscopic world's physics. Quantum Physics has been extensively used in electronics and Biochemistry. The modern-day world is full of Quantum Physics applications in various forms. It is like a box containing various gems like information about our Universe, atomic particles, forces, and much more about how everything works in the Universe. This article will help you know about different terms related to quantum physics for beginners to get a good head start on the subject.

The book will show you why quantum field theory is an important foundation for many other fields, including particle physics, condensed matter physics, and material science. You will also learn how these two disciplines are unified into a single study area known as Quantum Field Theory.

You will learn about all kinds of useful applications of quantum mechanics such as atomic clocks that can measure time much more accurately than previous methods, photovoltaic cells that work better in freezing temperatures, superconductors that allow electrical current to flow without resistance, and systems that can make any chemical reactions run more efficiently through chemistry controlled by brain waves. You will learn about the fundamental rules governing atomic structure that have led researchers to develop many materials such as graphite and diamond, whose properties are all related to quantum field theory. There will be many examples from engineering, biology, chemistry, physics, medicine, zoology, sociology, forensics, sociology, and economics where they are being used daily. You will also learn about many modern technologies that use the concepts presented in this book, including smartphones with fingerprint security measures based on quantum mechanics for highly sensitive data encryption.

In most people's eyes, science is hard to understand and quite boring to learn. However, Quantum Physics for Beginners will help you make an inspiring discovery. Quantum Physics seems like a complex topic at first glance, but it is not as complicated as other theories such as Newton's Laws of Motion. This guide has been written to make it very easy to understand quantum physics. You are not required to be a genius or a mathematics professional to know what quantum physics is all about.

Quantum Physics for Beginners is a guided textbook that will help you understand quantum mechanics basics from scratch and teach you the principles. When attending scientific conferences or Quantum Physics lectures, you will know exactly what they are talking about. You will learn about quantum physics' weird phenomena, such as entering two places at once, seeing things from two perspectives at once, and many more. These strange phenomena are often referred to as "spooky action at a distance" by Albert Einstein.

This guide explains Quantum Physics theories that are proven to be true. It does not take you long to understand these possibly confusing theories; they have been explained in simple terms without complex mathematical equations or molecular diagrams. Also, the basic principles have been explained in a step-by-step fashion. Even though pursuing further research doesn't interest you, you will clearly understand what quantum physics means.

Introduction

The book is for people new to quantum physics and old hands who want to learn more about these theories and find out more about the microscopic world around us. The book does not delve into theoretical topics such as dualism or intelligent life forms. This book aims to explain quantum physics principles without technical jargon thoroughly. What Is Quantum Physics?

For the vast majority of people, the term "quantum physics" is closer to "rocket science" than it is to "the wonders of the universe." And that's a genuine pity. Most of you might think of tedious formulae and explanations when thinking of physics, but both "traditional" physics and quantum physics are the sciences that hold the secrets to the Universe: The whys and the hows of how the entire cosmos works.

No matter what you do for a living, quantum physics will bring a whole new perspective into your life on so many things that it is impossible to ignore. How could you, when you know quantum physics is at the foundation of what you are, in the background of your fate spinning your life, and at the core of your way of "functioning" as an intelligent being of the Universe?

Almost borderline between science and spirituality, quantum physics might finally explain the unexplainable and help us transgress the borders of thinking, limiting us and bringing us closer to the essence of the world.

It comes from the fundamental theories that define this science. We invite you to discover the beauty of a study discipline that has been long considered a mystery and an impossible topic simultaneously, step by step.

Let's dive in and uncover the basics of quantum physics!

Chapter 1
What Is Quantum Physics and Quantum Mechanics?

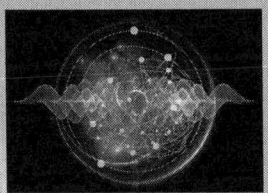

What is Quantum Physics?

Quantum physics is a branch of physics that deals with physical phenomena at microscopic scales. It specifically describes the behavior of objects on the smallest scales, i.e., distances and times shorter than atoms. Quantum physics has revealed a whole new aspect of reality, not directly observable with the human eye and senses alone. This new universe is one of radical uncertainty and pure energy. It contradicts our common sense, yet it is the most comprehensive and successful theory in all of science. Modern technology, such as computers, lasers, and transistors, has been based on quantum physics since it was first discovered.

How Does Quantum Physics Differ from Classical Physics?

Quantum mechanics differs fundamentally from classical physics. For instance:

- Rendering to Newton's first law of motion, an object should have a definite position and momentum at all times; however, according to Heisenberg's uncertainty principle, sub-atomic particles such as electrons do not have a well-defined position or momentum.
- A classical object has a definite mass. Electrons have a variable, and not a well-defined, mass.
- A classical object has a definite velocity. Electrons do not have a well-defined velocity; they just have a probability of being at various locations in space.
- In quantum mechanics, the properties of an object are represented by mathematical objects called wave functions. The uncertainty principle implies that even if you know all the components of a particle's wave function, you still don't know its location and momentum accurately. The position and momentum of an object are not determined until it is observed (so an observer becomes part of the system).

The uncertainty principle and wave functions are concepts that seem absurd. Many people became very confused by these concepts in the first few months. However, if you think about the implications of these ideas, it becomes easier to understand.

In this book, we will begin our study of quantum physics by finding out what it is. Before we can begin to understand how quantum physics works, however, we must first know just what we are studying to think about it intelligently. So, let us first consider some essential characteristics of quantum physics which will help us get acquainted with it and better understand its essence.

What Is Quantum Physics and Quantum Mechanics?

Some Essential Characteristics of Quantum Physics

1. Atoms, electrons, photons, etc., are not actually particles; they are "quanta" of various things (energy, mass, etc.).
2. The quantum physical world is very different from our "everyday" world. It is a strange universe where particles can be in two places at once, and objects can disappear and reappear in a new place without having traveled there in between.
3. Nothing happens in the quantum physical world unless somebody or something observes it. In other words, even if an event occurs (say, a photon emitted by an atom), it will not happen until there is an observer present to measure the event.
4. Time is not a constant quantity in the quantum physical world but comes in discrete jumps. One second can be equal to one hour in "quantum time."
5. There is such a thing as "negative" energy in the quantum physical world, which has the same magnitude as positive energy. This negative energy must be added to the positive energy to find the total energy of a particle or system (a fundamental law called "conservation of energy").

What is Quantum Mechanics?

Quantum mechanics is the branch of very small, related physics. It results in what might seem like very strange assumptions regarding the physical universe. Most of the classical mechanic's equations, which explain how objects move at ordinary sizes and speeds, cease to be useful on the scale of atoms and electrons. Through classical mechanics, objects reside in a particular location at a given time. However, through quantum mechanics, artifacts exist instead in a cloud of probability; they have a certain chance to be at point A, another chance to be at point B, and so on.

Basic theory in physics is quantum mechanics (QM; also known as quantum physics, quantum theory, wave mechanical model, and matrix mechanics), including quantum field theory. It explains advanced, atomic-scale properties of nature.

Classical physics explains many facets of nature at explaining physics that existed before relativity theory and quantum mechanics. Quantum mechanics also describes the existence dimensions at a small scale (atomic and subatomic).

Most theories in classical physics can be derived from quantum mechanics as a valid approximation on a large (macroscopic) scale. Quantum mechanics differs from classical physics. The energy, momentum, angular momentum, and other quantities of a bound system are restricted to discrete values (quantization). Objects have characteristics of both particles and waves (wave-particle duality).

Quantum mechanics emerged slowly, from hypotheses to clarify findings that could not be reconciled with classical physics, such as the response of Max Planck in 1900 to the question of black-body radiation and the correlation between energy and frequency in Albert Einstein's 1905 paper explaining the photoelectric effect. In the mid-1920s, Erwin Schrödinger, Werner Heisenberg, Max Born, and others profoundly re-conceived early quantum theory. The modern theory is formulated in various mathematical formalisms which are specially developed. In one, a mathematical function, the wave function, provides information on the probability amplitude of a particle's energy, momentum, and other physical properties.

What Is Quantum Physics and Quantum Mechanics?

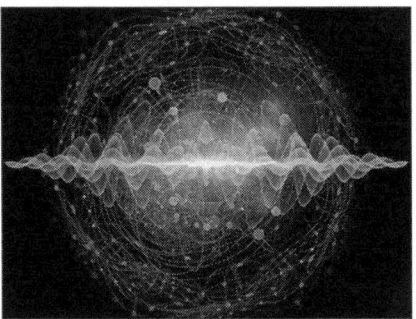

Quantum mechanics research is concerned with matter and light's atomic and subatomic scale behavior. It attempts to explain and consider the properties of molecules and atoms and their constituents — electrons, protons, neutrons, and other more abstract particles such as quarks and gluons. Such features include particle interactions with each other and with electromagnetic radiation (i.e., light, X-rays, and gamma rays).

Atomic-scale behavior of matter and radiation always appears unusual, and therefore the implications of quantum theory are difficult to grasp and believe. Its concepts are often at odds with common sense notions derived from everyday world observations. However, there's no reason why the atomic world's actions would adhere to the familiar, large-scale environment. It is important to note that quantum mechanics is a branch of physics and that the physics business is to explain and account for how the world — both on a large scale and a small scale — actually is, and not how one imagines or wants it.

There are many explanations for why the study of quantum mechanics is rewarding. Firstly, it demonstrates the physics' basic methodology. Second, it was highly effective in achieving accurate results in almost every case it was applied to. However, there is an interesting paradox here. Despite quantum mechanics' very overwhelming practical success, the subject's foundations contain unresolved issues — especially questions about the essence of measurement. An important feature of quantum mechanics is that the measurement of a system without disrupting it is usually difficult, except in principle; the precise essence of this disruption and the exact point at which it occurs is unclear and contentious. So, quantum mechanics attracted some of the 20th century's ablest physicists, and they created what is probably the period's finest intellectual edifice.

Both radiation and matter have particle and wave characteristics at a fundamental point. Scientists' gradual realization that radiation has particle-like properties and that matter has wavelike properties provided the development of quantum mechanics. Most 18th-century physicists, inspired by Newton, assumed that light consisted of particles called corpuscles. From around 1800, evidence for a wave theory of light accumulated. About this time, Thomas Young demonstrated that the two emerging beams interfere if monochromatic light passes through a pair of slits so that a fringe pattern of alternately bright and dark bands appears on a screen. A wave theory of light readily describes the bands.

According to the theory, when the crests (and troughs) of the waves from the two slits enter the screen together, a bright band is formed; a dark band is generated when the crest of one wave arrives at the very same time as the trough of the other, and then the effects of the two light beams cancel. Starting in 1815, a series of experiments by Augustin-Jean Fresnel of France and others showed that, when a parallel light beam passes through a neat single slit, the emerging beam is then no longer parallel, but it starts to diverge; this phenomenon is known as diffraction; because of the wavelength of the light and the configuration of the system (i.e., the spacing and width of the slits and the distance from the slits to the screen), the wave theory can be used to measure the predicted pattern in each case; the theory agrees with the experimental data precisely.

Chapter 2
Three Principles of Transformation

If you're curious about attraction, motion, and energy, you might be interested in some of the fundamental principles in quantum physics. The four laws of physics are not enough to describe all aspects of reality. Quantum Physics is needed for substances at a very small scale like atoms or photons. There are three laws that every quantum physicist knows about:

1. Superposition Principle — the wave-like nature of matter implies that a particle is always in more than one state or place at once until it's observed or measured.

2. Observer Effect — The act itself of measuring something can significantly change its properties, as well as our understanding and knowledge of what would have happened without the measurement being taken.

3. Wave-particle Duality — Matter can behave as either particles or waves.

Combined, they're called the Copenhagen Interpretation, where a particle is in many states, and a wave has an underlying particle nature.

For example, if you take a rock and drop it, the rock will fall to the ground. In Quantum Physics, you could explain why the rock falls by looking at each atom of table one at a time and seeing where any atom nears an atom of the rock; it could form a bond with that atom and start drawing energy toward it and building up more potential energy between those two atoms than other atoms that are further away from that pair. So, although it's not a direct cause and effect relationship between the atom you're looking at and the rock, there is a cause and effect connection that's formed from the billions of atoms in both objects.

Bohr's model of the hydrogen atom can be visualized as a tiny solar system. The sun is like the proton at its center orbited by one orbiting electron. To understand why this magnet-like field attracts the orbiting electron like gravity attracts our moon to earth, this model requires Quantum Physics because it needs to describe both natures at its smallest scale. It also needs to describe all aspects of reality at its largest scale.

The electron is always moving, but it's also in an orbit around the proton like you're always moving. But you're not really in orbit from moment to moment. You're just moving, but the earth is spinning faster than you're moving, so you can feel like you're in an orbit. So, what we have is a particle-like aspect with an underlying wave-like aspect of electrons and protons and orbiting electrons, which are both particles or waves, depending on if it's being observed or not.

So, when you take a hydrogen atom and pass an electric current through the wire to the proton at the center of that solar system-shaped atom, it gets hot because hot electrons near cold particles repel each other. For example, if you have a tub of water with warm water near the top and cold water near the bottom, there's not much cohesive energy between the warm and cold water. But if you mix warm and cold water into a spray bottle and shake it up, suddenly, there's a lot more cohesive energy between the waves because they are at higher frequencies. Suppose everything is hot all around you, like in an oven. In that case, that causes your body to go from being hot towards being very cold because hot waves will naturally repel each other, just like hot electrons will repel each other.

So, we have a proton with a positive charge at the center of an atom that's attracting an orbiting electron with its negative charge. Increasing the potential energy (like increasing the voltage) that wire is putting out attracts more electrons to orbit closer to this proton. At some point, if you increase it enough, it'll become unstable and either spit one or many electrons out of the atom-like shooting stars or fireballs. This can be hazardous to us and other things around us if they get hit with one of those free electrons flying through space at extremely high speeds when they leave the hydrogen atom because it's part of their momentum.

The wave-particle duality brings us to the Uncertainty Principle — the more you know about a particle, the less you know about its wave aspect and vice versa. This is because if you observe it with a microscope or any instrument, your observation will change its state from something that can behave as either wave or a particle.

The Uncertainty Principle is based on probability conservation. If we just look at tiny pieces of those objects without observing, information from different parts of those tiny pieces is undefined and can be anything we want because it's just probability.

If we observe a particle, it behaves as either waves or particles because the act of observation changes its state. As far as reality is concerned, it's only a wave if we leave it alone and don't measure it or observe it. Because of the Uncertainty Principle, nothing can be said about its location because observing where the electron is will change its momentum.

So, when I tell you that electrons are going to land in one place and not another, I'm only giving you extremely probable information about what could happen until you take an actual measurement.

That's why I say there's nothing in space because we're only giving you information about the probability of what could happen.

The basis of Quantum Physics is that anything that can happen will happen, but we don't know what will happen until an actual measurement is made. If we measure an electron on one sheet of paper and not the other, it will land on that piece of paper instead of the other one. But looking to see if your electron went to one or the other before you actually make a measurement didn't change its momentum as far as reality is concerned, so when you ask if it was going to hit either piece of paper, it was both pieces of paper.

As far as reality is concerned, the electron didn't land on either piece of paper until you went and looked to see if it landed there. So, for practical purposes, we can say that it lands on one side or the other with equal probability until we actually make a measurement.

But before we make that measurement, we have to consider everything else in the universe. If you're not alone in this universe (and you're not, so consider all those people around you), then they also have electrons flying around inside them with the same momentum and direction as yours.

All of those electrons are entangled with each other as well.

When you measure your electron going in one direction, the electrons in other people's bodies will also have their probabilities changed to reflect the fact that there's now twice as much momentum in that direction.

Suppose we're giving you information about probability instead of actual reality. In that case, it doesn't matter if it is a single particle or an ocean full of those particles. You can't predict everything about them because reality is based on probability, not a certainty.

If I give you a ball and tell you it's got an even chance of being red or blue, that means its color will probably be red or blue before you open the box. If you gave me a ball like that and asked about its color, I could tell you with equal probability that it was red or blue without actually looking inside the box.

You can't predict which one will be, but you can still answer any questions about the ball that don't rely on knowing its color.

The example above doesn't seem like a great quantum entanglement because we're talking about something we can see before we open the box, but that's because our human brains evolved to understand a world where reality is based on our senses.

The visible spectrum is just a tiny part of all possible forms of light, and the rules of probability are just as true for light as they are for electrons.

Most of the time, it's too hard to measure our light experiments, but there are plenty of things we can do on the quantum scale.

One classic example, for instance, is using two particles with a known spin pointing in opposite directions (either both up or both down) and measuring which way they're spinning.

If we measure one particle's spin and find its downward-pointing, we know that the other particle must have an upward-pointing spin. This is also true if we measure the first particle and find its upward-pointing spin.

In other words, by measuring one particle, we can instantly determine the other particle's spin, even though the two particles are very distant from each other and do not interact with each other except through the act of measurement.

Chapter 3
What Are Particles of Light?

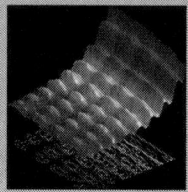

Understanding the Quantum Nature of Light

Physics is trying to interpret the laws related to movement and matter. However, quantum physics attempts to understand the behavior of the smallest particles and how they move. Such particles contain things like electrons, protons, and neutrons.

A. Quantum Physics. In Minute' Detail

In its emphasis on microscopic particles, quantum physics explains the particles that make up the tiny particles. The rules regulating macroscopic structures were wrong in setting precedents for smaller domains since the 20th century. The word "quantum" originates from the Latin term meaning "how much." It is used in physics to refer to the tiny matter and energy units whose action is predicted and observed in quantum physics.

Notably, even conditions that exist strongly and constantly, for instance, space and time, have values, although they seem to be of the smallest degree.

The atom's quantum model is even more complicated than we've seen before; instead of orbiting the nucleus like stars, electrons orbit in blurred, less defined, or cloud-like formation. Furthermore, the final configurations we have learned from the electron sequence (citing the number of electrons in the outer shell) are generally more like probabilities than hard and fast formation.

We bring this up when addressing the quantum nature of light to define the term quantum physics, so you can understand that its purpose is to show the numerical probability of the electron's place at any available time. Therefore, when the word is combined with "the essence of light," you should have a strong understanding of the general working principle.

B. Singular for Quantum Physics

Considering anything can conceivably influence the physical cycles that happen is one of a kind to quantum material science. For instance, in what is perceived as wave-molecule duality, light waves do indistinguishable particles, and these sections additionally show like waves. Put in another way, light has the two particles and waves' qualities, and either clarification can depict the activity of light.

In quantum tunneling, the issue can start with one area then onto the next without traveling through the mutual space. This offers a route to an advanced application where data can immediately go over great separations. We find that a great deal of the universe can be spoken to as a progression of probabilities through quantum material science.

There is a wide range of fields of quantum material science. The one that shines, especially on the conduct of lights (photons), is known as Quantum Optics. By investigating Quantum Optics, you will find that individual photon (light shafts) development examples legitimately influence the warm light.

This contrasts with the more common study of light, Classical Optics, acquired by Sir Isaac Newton, where the light was spotted as though it had lone particle belongings, meaning that it journeyed in a straight line, returned from objects with which it emanated into contact and dispatched through objects with minimum resistance.

C. Photons

To more readily comprehend what is suggested when the term photon is utilized, let us direct our concentration toward the Photon Theory of Light. A photon is a watchful group (or quantum) of electromagnetic (or light) vitality in this specific sense.

Existing in a vacuum and steady movement, photons have a consistent speed of light for all eyewitnesses. It occurs at the vacuum speed of light (more by and large alluded to as the speed of light), which is utilized.

$C = 2.998 \times 10^8$ m/s long.

Fixated on the Photon Theory of Light, the principle attributes of photons are as per the following:

- They prop up at a consistent speed, $c = 2.9979 \times 10^8$ m/s (light speed) in free space.
- They are known to have zero mass and zero rest vitality.
- They convey vitality and energy, which relate with the recurrence nu and frequency lambda, (and p, force) of the electromagnetic wave by $E = h\nu$ for and $p = h/\lambda$
- They can be destroyed or created when radiation is absorbed or emitted.
- They have the ability to have particle-like interactions, for example, electron impacts and other moment pieces.

D. Quantum Optics. Basic Understanding

To comprehend the quantum properties of light better, it may be useful to apply a portion of the related cycles (retention, emanation, and invigorated outflow) to the Laser, as this is one of the most notable utilization of quantum optics. Generally, these equivalent three attributes might be summed up to other light sources in differing degrees.

Electronic advances are normally the kinds of advances that transmit or ingest noticeable light. Simply envision an electron moving between measured nuclear vitality levels to perceive how this functions.

What Are Particles of Light? 15

For the Laser to work productively, the invigorated outflow of light is significant. Animated light emanation is utilized to give the enhancement needed to perform imaging work appropriately.

The exceptional property known as cognizance is the consequence of the animated emanation measure. The normal upgrade triggers the emanation occasions that are liable to give the enhanced light. These are associated with the photons released in the ideal advanced arrangement where each photon has the last stage relationship.

This type of intelligibility (relative arrangement) is characterized in two different terms: worldly cognizance and spatial soundness. Both end up being extremely critical in the advancement of obstruction used to produce 3D images.

Note: Ordinary light isn't intelligent because it starts from autonomous iotas that transmit around 10–8 seconds in time scales. Although there might be some rationality level in chosen sources, for example, the mercury green line and other valuable ghostly sources' sprinkling, their consistency isn't about what is contained in the Laser.

Few Specific Characteristics That Are Singular to Laser Light Include:

1. Cognizance: This is the property where different laser shaft parts are identified with one another in a stage relationship. When kept up over a long enough period, impedance impacts can be seen or photographically recorded. Intelligence is the factor that makes the possibility of visualizations conceivable.

2. Monochromatic: Comprising of one frequency, the laser light begins from the invigorated outflow of a solitary scope of nuclear vitality levels.

3. Collimated: Since they need to go through the mirrors a few times at amazingly opposite edges, the ways influenced by the intensification are known for their capacity to rebound back between the laser hole's finishes reflected firmly. For this reason, the laser radiates

have been intended to be extremely tight and restricted in their capacity to expand.

E. Photon and Probability

There are two ways the likelihood can be applied to photons' activity: the likelihood can be utilized to quantify the conceivable number of photons in a specific state, or the likelihood can be utilized to gauge the opportunity that a solitary photon will be in a specific state.

Since the previous understanding opposes Newton's Energy Conservation Principle, the last translation is the most practical other option.

Following crafted by physicist Thomas Young during the 1800s with a twofold cut investigation, Paul Dirac (1902-1984) was a hypothetical British physicist and one of the key pioneers of quantum material science; talks about this standard in his refreshed variant.

Indeed, even before finding quantum material science, individuals realized that a connection between light waves and photons must have a real character. Notwithstanding, it was not so sure that the wave work gave data on the likelihood of a photon in a fixed situation rather than the absolute plausible number of photons in that area.

This is an important distinction and can be clarified in the following way. Assume we have a light emission comprised of numerous photons isolated into two pieces of equivalent force. If the bar is connected to a sensible measure of photons, half of the likely number of photons ought to be reasonable to every distribution. When the parts are made to meddle with one another, one photon in one segment ought to upset the other.

According to the old theory, the two photons would have to cancel out one another, or they would have to generate four photons. Any outcome would contradict the principle of energy conservation.

Thus, according to the new theory, because the photon only slightly affects both elements, the question of relating the wave function to the probabilities for one photon only becomes a non-issue.

Each photon can only cause interference in this system, preventing two photons' potential occurrences.

Light Waves

Light waves are not the same as water waves and sound waves in that nothing is compared to the vibrating medium (for example, the water, string, or air) in the models discussed before. To be sure, light waves are fit for going through void space, as is evident from how we can see the light discharged by the sun and stars. This property of light waves introduced a significant issue to researchers in the eighteenth and nineteenth centuries. Some inferred that space isn't unfilled yet loaded up with a hidden substance known as 'aether,' which was thought to help the swaying of light waves. In any case, this speculation ran into inconvenience when it was understood that the properties required to help the exceptionally high frequencies average of light couldn't be accommodated as the aether offers no protection from the development of objects (for example, the Earth in its circle) through it.

Around then, the physics of power and attraction was being created, and Maxwell had the option to show that it was contained in many conditions (now known as 'Maxwell's conditions'). He additionally demonstrated that one kind of answer for these conditions compares to the presence of waves that comprise wavering electric and attractive fields that can go through void space without requiring a medium. The speed these 'electromagnetic' waves travel is dictated by the critical constants of power and attraction. When this speed was determined, it was seen as indistinguishable from the deliberate speed of light. This drove legitimately to the possibility that light is an electromagnetic wave. It is now realized that this model likewise applies to a scope of other marvels, including radio waves, infrared radiation (warmth), and X-beams.

Interference

Direct proof that a wonder, for example, light, is a wave, is acquired from contemplating 'interference.' Interference is usually experienced when two influxes of a similar frequency are included. Interference is critical proof for light's wave properties, and no other traditional model can represent this impact. Assume, for instance, that we rather had two surges of old-style particles: the total number of particles would consistently approach the whole of the numbers in the two bars, and they could always be unable to offset each other in the manner that waves can.

The primary individual to watch and clarify obstruction was Thomas Young, who, around 1800, played out an investigation. The light goes through a restricted cut named O, after which it experiences a screen containing two cuts, A and B, lastly arrives at a third screen, S, where it is watched. The light arriving at the last screen can have gone by one of two courses — either by A or by B. In any case, the separations that went by the light waves following these two ways are not equivalent, so they don't generally show up on the screen in sync with one another. It follows from the conversation in the past passage that, at certain points on S, the weaves will strengthen one another, while at others, they will drop; therefore, an example consisting of a progression of light and dim groups is seen on the screen.

Light Quanta

In 1905, Albert Einstein (around then obscure to mainstream researchers) distributed three papers that revolutionarily affected the fate of physics. One of these identified with the marvel of 'Brownian movement,' in which dust grains in a fluid are believed to move indiscriminately when seen under a magnifying instrument: Einstein demonstrated this was because of them being shelled by the particles in the fluid and this knowledge is commonly perceived to establish the last proof of the presence of atoms. Another paper (the one for which he is generally commended) set out the relativity hypothesis, including the popular connection between mass and energy. However, we are worried about the third paper — for which he was granted the Nobel Prize for physics — which clarified the photoelectric impact dependent on Planck's quantum theory.

Einstein understood that if the energy in a light wave is conveyed in fixed quanta, then when light strikes a metal, one of these will move its energy to an electron. Subsequently, the energy conveyed by an electron will be equivalent to that conveyed by a light quantum, less a fixed sum required to expel the electron from the metal (known as the 'work'), and the shorter the frequency of the light, the higher will be the energy of the discharged electron. When estimations of the properties of the photoelectric impact were broken down on this premise, it was discovered that they were in concurrence with Einstein's theory and the estimation of Planck. What was consistently found from these estimations was equivalent to that obtained by Planck from his investigation of warmth radiation.

A significant extra perception was that, regardless of whether the power of the light is feeble, a few electrons are transmitted promptly, the light is turned on, suggesting that the entire quantum is immediately moved to an electron. This is exactly what might occur if light were made out of a flood of particles as opposed to a wave, so the quanta can be thought of as light particles, which are called 'photons.'

In this manner, we have proof from the impedance estimations that light is a wave, while the photoelectric impact shows that it has the properties of a flood of particles. This is what is known as 'wave-particle duality.' A few readers may expect, or if nothing else, trust that a book like this will disclose to them how light can be both a wave and a particle. In any case, such a clarification presumably doesn't exist. The wonders that display these quantum properties are not part of our consistent experience (even though it is a significant point of this book to show what their outcomes are) and can't be completely depicted using old-style classes, for example, waves or particles, which our brains have advanced to use.

Indeed, light and other quantum objects are once in a while totally wave-like nor completely particle-like, and the most proper model to use for the most part relies upon the experimental setting. When we play out an impedance try different things with exceptional light emission, we, for the most part, don't watch the behavior of the individual photons. We can speak to the light as a wave to a generally excellent approximation. Then again, when we distinguish a photon in the photoelectric impact, we can helpfully consider it a particle. These depictions are approximations in the two cases, and the light joins the two viewpoints to a more noteworthy or lesser degree. Endeavors to understand quantum questions all the more deeply have raised theoretical difficulties and prompted incredible philosophical discussions in the last hundred or so years.

Such contentions are not key to this book, which plans to investigate quantum physics outcomes for our ordinary experience. The second arrangement of marvels that prompted the quantum to hypothesize is known as the 'photoelectric impact.' Electrons are discharged when light strikes a perfect metal surface in a vacuum. These all convey a negative electric charge, so the flood of electrons comprises an electric flow. Applying a positive voltage to the metal plate can stop this current, and the littlest voltage that can do so gives a proportion of the energy conveyed by every electron. When such analyses are done, it is discovered that this electron energy is consistently the equivalent of the light of a given frequency. If the light is made more brilliant, more electrons are radiated. However, the energy conveyed by every individual electron is unaltered.

Chapter 4
What Is the Wave of Matter?

The way light, which is customarily thought of as a wave, has particle properties drove the French physicist Louis de Broglie to conjecture that different articles we generally consider as particles may have wave properties. In this way, a light emission, which is most generally envisioned as a surge of tiny slug-like particles, would, in certain conditions, act as though it were a wave. This extreme thought was first legitimately affirmed during the 1920s by Davidson and Germer: they passed an electron bar through a precious stone of graphite. They watched an obstruction design that was comparative on a fundamental level to that delivered when the light goes through many cuts.

As we saw, this property is vital to the proof for light being a wave, so this test is an immediate affirmation that this model can likewise be applied to electrons. Later on, comparative proof was found for the wave properties of heavier particles, such as neutrons. It is currently accepted that wave-particle duality is a universal property of a wide range of particles.

Indeed, even ordinary articles, for example, grains of sand, footballs, or motorcars, have wave properties, even though in these cases, the waves are undetectable generally — mostly because the important frequency is excessively little to be recognizable. Yet since old-style objects are made out of atoms, everyone has its related wave, and every one of these waves is consistently slashing and evolving.

We saw earlier that on account of light, the vibration recurrence of the wave is straightforwardly corresponding to the energy of the quantum. On account of matter waves, the recurrence is difficult to characterize and difficult to gauge legitimately. Rather, there is an association between the frequency of the wave and the energy of the article. The higher the particle power, the shorter the matter-wave frequency.

In old-style waves, there is continually something that is 'waving.' Along these lines in water waves, the water surface goes all over; in sound waves, the pneumatic stress sways, and in electromagnetic waves, the electric and attractive fields change. What is the identical amount on account of matter waves? The ordinary response to this inquiry is that no physical amount compares to this. We can ascertain the wave using quantum physics' thoughts and conditions, and we can use our outcomes to foresee the estimations of amounts that can be estimated tentatively. However, we can't legitimately watch the wave itself, so we need not characterize it genuinely and should not endeavor to do so. To stress this, we use the term 'wave work' instead of a wave, which underlines the point that it is a numerical capacity instead of a physical article.

Another significant specialized contrast between wave capacities and the traditional waves we talked about before is that while the old-style wave wavers at the wave's recurrence, in the matter-wave case, the wave work stays consistent in time. Notwithstanding, because not physical in itself, the wave work assumes a basic job in quantum physics to understand genuine physical circumstances. Right off the bat, if the electron is limited to a given district, the wave work forms standing waves like those discussed before; subsequently, the frequency and along these lines, the particle's energy takes on one of a lot of discrete quantized qualities. Furthermore, if we do trials to identify the nearness of the electron close to a specific point, we are bound to discover it in areas where the wave work is enormous than in ones where it is little. This thought was set on an increasingly quantitative premise by Max Born, whose standard expresses that the likelihood of finding the particle at a specific point is relative to the block of the extent of the wave work by then.

Atoms contain electrons bound to a little district of the room by the electric power pulling them to the core. We could expect the related wave capacities to frame a standing-wave example from what we said before. In a matter of seconds, we will see how this prompts an understanding of significant appropriate ties of particles. We start this conversation by considering a less complex system where we envision an electron to be limited to a little box.

Chapter 5
Principle of Uncertainty

"Anyone who is not surprised by quantum theory has not understood it," said Neils Bohr, a pivotal contributor to quantum mechanics theory.

Such is the beauty of entanglement theory; as more and more years go by, and as more and more scientists get their hands on it, they have to find new ways of explaining their shortcomings — the uncertainty principle is the most prominent.

Also called the Heisenberg uncertainty principle and the indeterminacy principle, the uncertainty principle states that there is no exact measurement of the position or velocity of an object in the quantum world. The concept of exactness has no place in this realm, not even in theory.

The uncertainty principle regards only tiny objects as immeasurable because it applies to the quantum world. For this reason, ordinary objects do not apply to the uncertainty principle. There is proof of this: anyone can find an exact measurement of a car because they can weigh it. It is ordinary and, therefore, large enough to be accurately pinpointed.

Even a category of tiny objects qualifies for more accurate

measurement than the uncertainty principle makes a room. These objects are ones whose velocity and position are equal to or greater than Planck's constant, 6.6×10^{-34} joule-second. Small items below Planck's constant apply to the uncertainty principle.

The uncertainty principle arose from classic wave-particle duality. Every particle has an accompanying wave. The more undulating that lock is, the more uncertain its measurement is. The more specific its accompanying particle's position, but indefinite its momentum.

Heisenberg also made another significant contribution to quantum mechanics in 1927. He argued that because matter behaves like waves, some properties, such as the position and speed of the electron, are "complementary," implying that there is a limit (related to the Planck constant) to how well the accuracy of and property can be understood. Under what would come to be called the "Heisenberg Theory of Uncertainty," some argued the more exactly the electron's position is determined, the less exactly its speed can be identified and vice versa. This uncertainty theory often applies to everyday objects but is not apparent since the lack of precision is too high.

Werner Heisenberg (1901-1976) was the theorists' prince, so disinterested in laboratory practice, he risked flunking his thesis at the University of Munich because he did not know how batteries worked. Fortunately for him and physics, he was also promoted. There were other difficult moments in his life. During the First World War, while his father was at the front as a soldier, the scarcity of food and fuel in the city meant that schools and universities were often forced to suspend classes. And in the summer of 1918, young Werner, weakened and malnourished, was forced together with other students to help the farmers on a Bavarian farm harvest.

With the end of the war, in the first years of the twenties, we find him in the shoes of the young prodigy: pianist of high level poured in the classical languages, skillful skier and alpinist, and mathematician of rank lent to the physics. During the old teacher Arnold Sommerfeld's lessons, he met another promising young man, Wolfgang Pauli, who would later become his closest collaborator and fiercest critic. In

1922, Sommerfeld took the 21-year-old Heisenberg to Göttingen. The beacon of European science attended a series of lectures dedicated to the nascent quantum atomic physics, given by Niels Bohr himself. On that occasion, the junior researcher, not intimidated, dared to counter some guru statements and challenge at the root of his theoretical model. However, after this first confrontation between the two, a long and fruitful collaboration was born, marked by mutual admiration.

From that moment, Heisenberg devoted his body to quantum mechanics' enigmas. In 1924, he spent some time in Copenhagen to work directly with Bohr on radiation emission and absorption problems. There, he learned to appreciate the "philosophical attitude" (in Pauli's words). He was frustrated by the difficulties of concrete Bohr's atomic model, with its orbits put in that way; who knows how the young man was convinced that there must be something wrong at the root. The more he thought about it, the more he thought those simple, almost circular orbits were a surplus, a purely intellectual construct. To get rid of them, he thought that the very idea of rotation was a Newtonian residue that had to be done.

The young Werner imposed himself a fierce doctrine: no model had to be based on classical physics (so no miniature solar systems, even if they are so cute to draw). The way to salvation was not intuition or aesthetics but mathematical rigor. Another of his conceptual digits was the renunciation of all entities (such as orbits, in fact) that could not be measured directly.

Measurable in the atoms were the spectral lines, the witness of photons' emission or absorption by the particles resulting from jumping between the electron levels. So, to those net, visible, and verifiable lines corresponding to the inaccessible subatomic world, Heisenberg turned his attention. To solve this diabolically complicated problem and find relief from hay fever, in 1925, he retired to Helgoland, a remote island in the North Sea.

His starting point was the so-called "correspondence principle," enunciated by Bohr. According to this, quantum laws had to be trans-

Principle of Uncertainty

formed without problems into the corresponding classical rules when applied to sufficiently large systems. But how big? Enough to allow to neglect the Planck constant h in the relative equations. A typical object of the atomic world has a mass equal to 10^{-27} kg; let's consider that a grain of dust barely visible to the naked eye can weigh 10^{-7} kg: very little, but it is still more significant by a factor of 100000000000000000000, that is 10^{20}, one followed by twenty zeros. So, the atmospheric dust is clearly in the domain of classical physics: it is a macroscopic object. Its motion is not affected by factors dependent on Planck's constant. The fundamental quantum laws apply naturally to the atomic and subatomic world phenomena. Simultaneously, it loses sense to use them to describe phenomena related to aggregates larger than atoms as the dimensions grow, and quantum physics gives way to the classical laws of Newton and Maxwell. The foundation of this principle (as we will repeat several times) is that the strange and unpublished quantum effects "correspond" directly to physics' classical concepts as you leave the atomic field to enter the macroscopic one.

Driven by Bohr's ideas, Heisenberg redefined the banalest notions of classical physics in a quantum field as the position and velocity of an electron. They were in correspondence with the Newtonian equivalents. But he soon realized that his reconciliation efforts between two worlds led to the birth of a new and bizarre "algebra of physics."

In school, we all learned the so-called commutative property of multiplication. Given any two numbers and b, their product does not change if we exchange them between them; in symbols: $a \times b = b \times a$. It is clear, for example, that $3 \times 4 = 4 \times 3 = 12$. However, in Heisenberg's time, abstract numerical systems in commutative property do not always apply. It is not said that $a \times b$ is equal to $b \times a$. Examples of non-commutative operations are also found in nature to think about it. A classic case is rotations and tilts (try to perform two different wheels on an object like a book, and you will find examples where the order they happen is essential).

Chapter 6
Differences between Quantum Physics and General Relativity

For the changing society of the early twentieth century, using only mathematics, incredible aspects of nature could be described as a pleasant surprise. Perhaps not that it was an unprecedented experience, but few expected such a revolutionary change in a paradigm that had lasted for centuries. Before Einstein, the Newtonian description of gravity was revered for its simplicity and universal validity.

Relativity was soon discovered to hold more surprises. In 1915, the same year of its presentation in public, the theory revealed the possible existence of astrophysics bodies. Physics with stellar masses do not let even light escape or black holes. These matter-eaters beasts immediately became the most seductive and enigmatic celebrities of relativity for a wide audience. Soon after, Einstein showed that his theory predicted the existence of gravitational waves, deformations of the cosmic tissue that move at the speed of light as if they were very fast space earthquakes. As if this were not enough, it was soon realized that a detailed scientific description of the universe's history was possible based on the equations of relativity.

Despite the interest of these last entries, for several decades of the last century, the study of black holes, gravitational waves, and cosmology were seen as nothing more than curious theoretical works in more conservative scientific circles. The reason was that it is difficult or impossible to make direct confirmations about these theories and that it is difficult to make precise relativistic calculations to compare them with indirect observations.

However, ingenious (and sometimes just lucky) experiments have accumulated vast indirect evidence. For example, in cosmology, the discovery of cosmic background radiation, predicted and described by the Big Bang cosmological model as remnant radiation from an early time when the universe was very hot, was an important building block in the consolidation of cosmology. This discovery was award-winning. Despite this, there were still doubts on whether the big bang model, based on general relativity, was correct to describe this radiation. Fortunately, a decade later, it was confirmed that radiation is not uniform throughout the cosmos and that the measurement of these inhomogeneities matches what the theory had predicted. Celebrated with the Nobel Prize in 2006, this discovery did not leave many options open. The cosmology described by relativity is the most appropriate description. A more detailed measurement of the universe's expansion soon led to the last greatest cosmological discovery: the universe grows faster and faster (Nobel Prize 2011). And (almost) everything can be perfectly adjusted from general relativity.

The confirmation of black holes and gravitational waves does not have a shorter history. For one thing, black holes have always represented a singular scientific nuisance. The fact that the theory indicates that the gravitational force at the heart of black holes is infinite indicates a serious problem: right there, the theory of general relativity is no longer valid. Thus, it was considered that black holes were a trick that mathematicians played on us for some time. However, theorists argued that, like stars made entirely of neutrons, black holes were corpses of stars larger than ours. The confirmation of the existence of neutron stars in the 1960s and indirect observations of the movement of stars around dark regions led to the certainty that there are many black holes in the universe and that they can have masses millions of times that of the Sun.

Additionally, black holes, while not the typical cosmic vacuum cleaners that we paint, absorb large amounts of matter from their cosmic neighborhood, creating a ring of incandescent radiation-emitting material called an accretion disk. This radiation predicted by the theory has been confirmed, especially in the center of galaxies like ours. In the Milky Way, the movement of a group of stars around the center of the galaxy, "chased" by astronomers since the 1990s, has revealed that there lives a relatively small and dark body with a mass of almost 3 million times that of the Sun and that it emits radiation according to predictions for a hole. As if this were not enough, in 2011, an accretion disk was observed with emission of X radiation, consistent with the predictions of black holes with masses of billions of times that of our star, absorbing material from a quasar.

The last series of indirect observations of black holes involves the gravitational waves. Although these can occur with any accelerated movement (even a clap produces them), according to the theory, only gravitational waves generated by violent cosmic events are capable of producing gravitational waves that we can detect on Earth, with sensors so sensitive as to measure deformations of the space of a thousandth the size of a proton or less. Gravitational-wave detection led to the award of the 2017 Nobel Prize. Supercomputers with sophisticated programs managed to show that the signals, according to relativity, were only consistent with the collision and mixing of two black holes with masses equivalent to a few dozen times that of the Sun. All at once, two of the most controversial predictions of relativity were proven just a couple of years ago! And the evidence continues to accumulate today to explore possible deviations from the predictions of Einstein's theory and find astronomical applications of the study of gravitational waves.

Not only gravitational waves and black holes are under the scrutiny of research. As we said, the cosmological model of the big bang can explain all observations of the dynamics of the universe under very simple assumptions on the geometry of space-time and an assumption on the content of the cosmos based on recent measurements, which is: 5% matter like that of our planet, 27% a type of matter called dark matter that does not emit light, and 68% of cosmic content is a form of energy nicknamed dark energy that causes the accelerated expansion that we observe.

The biggest mystery is that no one has the slightest idea what are dark matter and dark energy. It is nothing that we have been able so far to observe directly, although there is sufficient indirect evidence to affirm that there are such "substances" or something that has the same effects. However, some are convinced that we must slightly modify the basic equations of general relativity to understand the true nature of these dark entities. Others consider that only the immensely challenging search for the compatibility of general relativity and quantum mechanics will help us dissipate our doubts.

And this mix brings us today back to black holes. The inside of the black holes, being not observable, is totally unknown. All we know is that gravity must be so strong inside that it could affect the smallest particles, comparable to the effects of quantum forces that govern their behavior according to particle physics. If so, a form of quantum gravity may be manifested there, which we must theorize based on what we have verified in the last century.

But even for those least interested in the fundamentals of gravity, the theory of relativity today offers important modern tools. In addition to being relevant for the global positioning system (GPS), it is crucial in astronomical observations. The deflection of light due to its passage near galactic formations, stars, planets, etc., causes an apparent shift in the position of the stars and galaxies concerning the real one. But it is not the only effect. If there is a galaxy behind a very massive astrophysical body, the deflection of light beams emitted by the galaxy in all directions can be deflected towards us around the contour of the galaxy.

Astrophysical "nuisance." This effect is called gravitational lensing. Gravitational lenses not only allow us to characterize what is behind the observable objects that cause the deflection of light but also when they occur in regions where there are no visible obstacles, they show us properties of unobservable objects, such as black holes and dark matter formations, which have not yet been fully described.

Relativity, despite its age, remains a developing treasure whose questions and responses pose current challenges that are likely to become the basis for future discoveries and a paradigm shift like the one Einstein witnessed.

Quantum Mechanics and General Relativity Incompatibility

Physicists are trying to reconcile the two textbooks according to which science understands the world. Still, to date, there has surfaced no proven, palpable theory to bring the two worlds together and finally help us understand where we come from, where we are, and where we are going (because, at the end of the day, these are the fundamental questions both classical and quantum physics propose).

In classical physics (as drawn out by Einstein's general relativity principle), the reality is made out of 4 dimensions (also called the space-time continuum). In this paradigm, gravitational fields are continuous entities.

In quantum mechanics, however, fields are not continuous but discontinuous. They are not defined by the 4 dimensions but by "quanta." As such, concepts like the "gravitational field" are missing from the world of quantum physics, which is also the biggest bridge classical physicists and quantum researchers have to build between their points of view.

This is not just a matter of fancy definitions. The world of quantum mechanics and classical physics are incompatible because they describe reality in completely different ways, in different terms, and in different perspectives that do not meet at any point.

In classical physics, things happen for a reason. They happen

Differences between Quantum Physics and General Relativity 33

according to the old cause-and-effect dictum. Nothing happens randomly but because something else before it has caused it.

In quantum physics, scientists do not see reality in terms of cause and effect but in terms of particles jumping from one state to another based on probability rather than definite outcomes.

Why is reconciliation important, then, especially given that these two disciplines seem so different and at such a deep level?

Because reconciliation would also bring along complementarity relationships, where classical physics fails to give explanations of the microcosmos, quantum physics would succeed. And where quantum physics fails to make sense when it is blown up to macro objects (remember the cat that was both dead and alive?), classical physics would be able to breathe in some logic.

Chapter 7
The Schrodinger's Cat

As a basic theory, quantum mechanics should, in principle, apply to physical systems of any size, that is, not only limited to microscopic systems. It should provide a method for transitioning to classical macroscopic physics. The existence of quantum phenomena raises a question: how to explain the classic phenomena of macroscopic systems from the viewpoint of quantum mechanics. What can't be seen directly is how the superposition state in quantum mechanics can be applied to the macro world.

In 1954, in a letter to Max Bonn, Einstein raised the question of how to explain macroscopic objects' positioning from the perspective of quantum mechanics. He pointed out that the phenomenon of quantum mechanics is too small to explain this problem. Another example of this problem is the Schrödinger proposed Schrödinger's cat thought experiment. However, the experiment that did not occur in the real world and was only imagined can be introduced briefly before going into its details afterward in the book. It consists of putting a cat in a small box. Then, you expose it to a radioactive substance. When the substance decays, it triggers the Geiger counter device and therefore kills the cat by the poisonous substances released by the explosion that may take place. By a double approach, the radioactive substance's decay is controlled by the binary status of 'going to decay' and 'not going to decay,' which means simultaneously the superposition of two paradoxical states. In other terms, the cat is both alive and dead simultaneously.

It was not until about 1970 that people began to truly understand that the thought experiments mentioned above were not practical because they ignored the inevitable interaction with the surrounding environment. It turns out that the superimposed state is very susceptible to the surrounding environment. For example, in a double-slit experiment, the collision or emission of electrons or photons with air molecules can affect the phase relationship between the states critical to the formation of diffraction.

In quantum mechanics, this phenomenon is called quantum decoherence. It is caused by the interaction of the system's state with the effects of the surrounding environment. This interaction can be expressed as the entanglement of each system state with the state of the environment.

The result is that superposition is only effective when considering the entire system (i.e., experimental system + environmental system). If only the experimental system's state of the experimental system is considered in isolation, only the system's classic distribution is left. Quantum decoherence is the main way to explain macroscopic quantum mechanics' classical properties in quantum mechanics today. It is also the biggest obstacle to the realization of quantum computers. Multiple quantum states need to remain superimposed for as long as possible in a quantum computer. Short decoherence time is a huge technical problem.

Chapter 8
The Building Blocks of Matter and Wave-Particle Duality

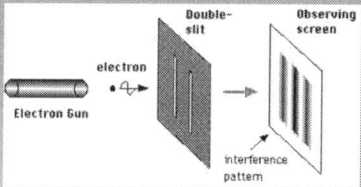

Despite the typical wave behavior observed in the experiment, light still has particles' properties. Firstly, as we already know, it is divided into quanta called photons. Secondly, it can leave shadows and patterns resulting from holes in the wall. Additionally, if only one slot is left open during the experiment, one neat band is formed opposite this slit, resulting from the particles' flux. How do we understand this dual nature of light, and how can we describe it? Why does light act like waves in one case and like particles in another?

At first, scientists tried to explain where the waves come from using the water analogy: light is a collection of particles, just like a water body is a collection of water molecules, and a set of particles, like a large number of water molecules in water bodies, can form waves. Therefore, each quantum, each photon (single unit of light), must be a particle. It was easy to test in a new version of the two-slit experiment, but this experiment did not confirm expectations! When the photons were fired one by one (for example, one per minute) towards the slit, each one appeared on the second wall, not in front of either slit, but randomly in one of those scattered places where interference bands had appeared in the standard version of the experiment! The wall retained a visual trace of the light particle (this was not just a wall, but a unique screen that contained all the light traces), and over time, with each subsequent photon, an interference pattern was increasingly evident.

Why do the individual photons not appear on the screen directly in front of the slits? Why don't two stripes form on the screen?

This behavior of every single photon was utterly unexpected and incomprehensible. The individual photons could not interact with any other particle because the photons were fired one at a time with a gap in between, much more spaced apart than light quanta usually travel. However, the final position of each photon at the screen was the result of interference. At the same time, upon getting to the screen, every single photon still left a point trace, just as would be expected of a particle.

These results cannot be explained in terms of the reality known to us. It seems evident that those photons could somehow move between the state of either particles or waves. No one had ever encountered anything like this before. The facts of quantum reality that we will discuss are no less weird, but these are real facts of the microcosmic world since all of these are the results of observations and controlled experiments. Is it so unusual that this field is difficult to understand since it consists of a range of entirely new phenomena from the microcosmic level of reality that cannot be explained by the notions customary to us and even contradict them? Our perceptions of what is generally possible turned out not to be final since, until then, we had been dealing only with our macro world with its simpler laws and interconnections between the facts. The new facts related to the micro world, which have no analogs in the macro world, seem strange and often even unbelievable. In the language of science, this refers to the difference between quantum physics and regular classical (Newtonian) physics known to a certain extent to each of us from school and everyday experience.

The fact that individual photons exhibit properties of both parti-

where each electron came from and, therefore, predict where they might go.

That guess is based on probability because each electron acts like a wave with two different destinations (one for each slit). Still, electrons are particles, so they aren't allowed to go through both slits at once, and so they will act like a particle (a particle is something that never acts like a wave).

Feynman came up with a way to save electrons from going through both slits by looking at each electron as an event. By doing this, you can break down the wave function into smaller components that are all causally related (i.e., the electron will end up in one of these spots purely because it has been broken down into smaller pieces).

Since there is no such thing as time in quantum mechanics, there is no way for us to know if an electron has gone through a slit or which slit (and how could we?). All we can do is measure what its path is once it's on our screen.

This makes sense because what we see in our universe, at the macroscopic level, is mostly based on classical physics, which is the physics of big things like objects and people. If you try to apply quantum mechanics to these giant objects (like a hand that you place through a slit), then your results will make no sense and have no explanation.

The von Neumann (1956) formulation of quantum mechanics explained that all processes in the universe could be broken down into elementary units or atomic components. These atomic components are called atoms or elementary particles, which make up everything we see around us. This is also known as wave-particle duality.

What is Wave Interference?

If two rocks are thrown together into a pond, each creates waves in the form of ripples that move away from the location where the rock hits. When the duck floats on the shallow surface, the impact of both

The Building Blocks of Matter and Wave-Particle Duality   37

ference bands had appeared in the standard version of the experiment! The wall retained a visual trace of the light particle (this was not just a wall, but a unique screen that contained all the light traces), and over time, with each subsequent photon, an interference pattern was increasingly evident.

Why do the individual photons not appear on the screen directly in front of the slits? Why don't two stripes form on the screen?

This behavior of every single photon was utterly unexpected and incomprehensible. The individual photons could not interact with any other particle because the photons were fired one at a time with a gap in between, much more spaced apart than light quanta usually travel. However, the final position of each photon at the screen was the result of interference. At the same time, upon getting to the screen, every single photon still left a point trace, just as would be expected of a particle.

These results cannot be explained in terms of the reality known to us. It seems evident that those photons could somehow move between the state of either particles or waves. No one had ever encountered anything like this before. The facts of quantum reality that we will discuss are no less weird, but these are real facts of the microcosmic world since all of these are the results of observations and controlled experiments. Is it so unusual that this field is difficult to understand since it consists of a range of entirely new phenomena from the microcosmic level of reality that cannot be explained by the notions customary to us and even contradict them? Our perceptions of what is generally possible turned out not to be final since, until then, we had been dealing only with our macro world with its simpler laws and interconnections between the facts. The new facts related to the micro world, which have no analogs in the macro world, seem strange and often even unbelievable. In the language of science, this refers to the difference between quantum physics and regular classical (Newtonian) physics known to a certain extent to each of us from school and everyday experience.

The fact that individual photons exhibit properties of both particles and waves proves that our strict division of reality into particles and waves is incorrect. Things aren't as simple as we thought. It turns out that particles and waves are concepts that may relate to the same phenomenon (for example, waves of radiation and photons of radiation. The term "photon" is used concerning not only visible light). But what about the solid matter, which consists of particles?

Solid matter consists of atoms. An atom consists of particles: electrons, protons, and neutrons (the latter two consist of even smaller particles, namely, quarks). Further experiments showed wave properties displayed by electrons, neutrons, and even entire atoms and molecules! Everything that makes up what seems to be solid matter acts like waves as well! Each particle of matter can "blur" its position. This dual nature of the whole reality is called wave-particle duality. Everything is made of particle waves.

But why does everything sometimes acts like particles and sometimes like waves? This is hard for people to imagine, even today.

Chapter 9
Quantum Possibilities and Waves

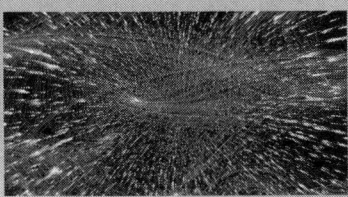

Quantum physics can be a complicated subject. It's been said that quantum mechanics is the physics of possibilities. To reach quantum conclusions, you need to ask the "what if" questions and follow how a particle will behave. Quantum also has the concept of uncertainty, which is an odd thing considering that these are particles in some sense knowable. To enjoy the simplicity of quantum physics, let's start with a simple example.

If you have a particle with an even mass number and you want to find the location of that particle, you can perform two experiments. In experiment one, the particle doesn't move, and in experiment two, it moves towards its right. We can begin with these experiments and then make assumptions about where the particles will end up based on probabilities or uncertainty.

The first experiment points a stream of subatomic particles (electrons) at a sheet of metal with two slits in it, then counts how many particles go through each slit. In the second experiment, we will have slits, but this time with a magnet to move the particles to the right. The reason for this is because we know that the magnet will be attracted by a positive charge and pushed away by a negative one, so if both of our particles have a negative charge, we know that it will be pushed towards its right.

The results of experiment one would lead us to believe that if you drop an even number of molecules (particles), half of them will go through the hole on the left and half through the hole on the right. In stark contrast, in experiment two, it would seem that every single particle goes out only one hole.

As physicists, we can assume that half of the particles go through both holes, and the other half is blocked. We then assume that every particle only wants to go through one hole and one hole. We count more than half of the particles going through a single hole because some want to go out of both holes, but there's no way to know exactly how many went out of both holes, which will contribute to uncertainty. It would be even more uncertain if we used larger or smaller particles because they are harder or easier to detect (remember, we're talking about electrons). In smaller particles, it's easier to miss the mark.

Now that we have a better idea of a particle's probable location let's look at another example. Say you want to predict where an electron will appear on a screen when someone flashes a light at it. The experiment goes like this: you fire electrons at a screen that has two vertical slits in it (one for each slit you fired electrons out).

There is zero uncertainty in its starting position because you fired the electron from one point. You will see that each electron behaves differently and goes through different holes, but every single time they will go through one hole and never both holes.

After the experiment is over, you go back and look at your data. You notice that the electrons act in a strange way. Not only do they not behave like waves, but they appear to be able to know in advance where the second slit is.

Quantum mechanics looks at where electrons will appear on a screen based on probability. We can assume that every electron wants to go through both slits, but they can't be in two places at once, so they must appear by chance. We can assume that the electrons aren't in a wave state, so they appear in a certain place out of all possible places.

Because there are two slits, we know there will be interference, and it will look like the electrons are waves. We also know that there is zero uncertainty on its starting position because we know where each electron came from. Also, because there was only one electron fired at a time, there is no uncertainty regarding how many electrons went through which slit.

In this experiment, we can assume that each electron will go through one of the two holes. We then devise a mathematical equation to show this. In the quantum world, things are bound to happen because of probability, so the probability wave theory comes in. A probability wave is like an invisible force that tells the particle where and when it should appear on a screen.

This concept is called the Copenhagen Interpretation, but it has been revised over time as people have come up with new methods to test quantum mechanics. The first iteration of the Copenhagen interpretation came from Niels Bohr in 1927 and was named after his hometown of Copenhagen, Denmark (it's also known as the cradle of quantum physics).

The Copenhagen interpretation tells us that, at the very core of everything, there is a particle that has a nucleus made up of protons and neutrons, and it has electrons orbiting it (that might be tense due to gravity). These electrons are all over the place in a haze of uncertainty.

For an electron to be identified as an electron, the particle will have to go through one hole and not both. For this to happen, you need an observer or detector. You can imagine that if you had no detectors (e.g., eyes), then you would see a probability wave interference pattern on your screen with no definite measurement results.

Now, let's imagine that you have a detector. The observer, or the detector, has to look at the screen for a particle to be identified as an electron. By looking at the experiment results and comparing them with what data you already have, you can make an educated guess on where each electron came from and, therefore, predict where they might go.

That guess is based on probability because each electron acts like a wave with two different destinations (one for each slit). Still, electrons are particles, so they aren't allowed to go through both slits at once, and so they will act like a particle (a particle is something that never acts like a wave).

Feynman came up with a way to save electrons from going through both slits by looking at each electron as an event. By doing this, you can break down the wave function into smaller components that are all causally related (i.e., the electron will end up in one of these spots purely because it has been broken down into smaller pieces).

Since there is no such thing as time in quantum mechanics, there is no way for us to know if an electron has gone through a slit or which slit (and how could we?). All we can do is measure what its path is once it's on our screen.

This makes sense because what we see in our universe, at the macroscopic level, is mostly based on classical physics, which is the physics of big things like objects and people. If you try to apply quantum mechanics to these giant objects (like a hand that you place through a slit), then your results will make no sense and have no explanation.

The von Neumann (1956) formulation of quantum mechanics explained that all processes in the universe could be broken down into elementary units or atomic components. These atomic components are called atoms or elementary particles, which make up everything we see around us. This is also known as wave-particle duality.

What is Wave Interference?

If two rocks are thrown together into a pond, each creates waves in the form of ripples that move away from the location where the rock hits. When the duck floats on the shallow surface, the impact of both waves can be felt. At certain points in the pool, the behavior of the two waves reinforce each other, creating a broad up-and-down oscillation (the duck has a crazy ride). This reinforcing effect is called positive interference. But, at certain places in the bay, the behavior of the two waves may be reversed, causing no up-and-down oscillation. (The duck is still there.) This canceling effect is called destructive interference.

Chapter 10
"Dark Body" Spectrum Understanding the Curve of the Black Body

Heated Bodies Radiate

For now, we will go to another puzzle that despises scientists as the new century turns (1900): how do warm corpses begin? There was a complete understanding of the system in question — the heat was known to cause particles and iotas to vibrate vigorously, and particles and molecules were proven examples of electrical charges. (Obviously, Newton was in good shape.) From the investigation of Hertz and others, Maxwell's prospects for light-emitting diversion cases have been confirmed, in any case. It was known from Maxwell's circumstances that the radiation traveled at the speed of light. In this case, it was understood that the light itself and the warm rays associated with the field were actually electric waves. At the time, the picture was that when the body was heated, the subsequent vibrations on the sub-atomic and nuclear scale inevitably removed. Acknowledging at the time that Maxwell's idea of electromagnetism, which is the most efficient in the physical world, was legitimate and at the sub-atomic level, these desirable costs would have passed, perhaps emitting warmth and visible light.

How Is Radiation Absorbed?

What is meant by the expression "dark body"? The fact of the matter is that the radiation from the hot body depends in some way on the body being heated. We should immediately support and consider how different materials store radiation to see this with great success. A few, such as glass, seem to bring light in any way — light passes directly. With a shiny metal surface, light is not included.

It may be visible. Dark materials such as ash, light, and heat are completely compressed, and the equipment is warm.

How can we understand these various processes, such as light waves that adapt to changes in applications, making these charges affect and store energy from radiation? Thanks to the glass, it is clear that this is not happening, somehow very little. Why not? Full understanding of why it requires quantum equipment; however, the general idea is as follows: there are costs — electrons — in the glass that can fluctuate in the light of the affected electric field outside, but these charges are firmly attached to the molecules and can only fluctuate in certain waves. (In quantum artisans, these charging vehicles occur when the electron moves from one circle to the next. Noticeable, so there is no recurrence with a small wave, and the vigor is obtained from now on. That's why glass is perfect for windows! Duh). However, the glass is not clear on certain waves outside the visible distance (as a rule, both infrared and light). These are waves where the distribution of electric charges on iotas or bonds can often fluctuate.

How can we understand the thinking of light through metal? A small metal has allowed electrons to travel at all power. This is what makes iron into metal: it conducts energy and heat effectively; the flow of these transparent electrons actually transmits both. (All things considered, a little warmth is transferred to the vibrations.) But metals are irresistible because they sparkle — why is that? Once again, those free electrons are pressed into larger segments (compared to particles) by the electric field of the approaching light waves, and this stimulating flow comes from the electric field, much like the flow in a talking radio wire. The radiation is the reflected light. With a shiny metal surface, a slight glow of glamor is combined with the heat; it is automatically renewed and visible.

At the moment, what about looking at something that shines a light: there is no transmission and no display. We are approaching the best end with ashes.

Like steel, it will lead to the flow of electricity; however, not even a successful approach. There are unconnected electrons, which can travel at all energies but continue to hold objects — they have a short-term meaning. When they knocked, they caused a commotion, like the balls hitting the guards on a pinball machine, so they gave off a strong force in the heat. Apart from the fact that the electrons in the ash have a shorter duration than that of the noble metal, they are clearly compared to the electrons bound by iotas (as in glass) to accelerate and gain energy in the electric field. In this way, they are powerful mediators in moving energy from light waves to heat.

Absorption and Excretion

After seeing how the ashes can get into the rays and transfer energy to the heat, shouldn't that be said about talking? For what reason does it transfer when it is heated? The similarity of the pinball machine is still acceptable: think now of the pinball machine where the boundaries are, and so on. Vigorously vibrate because they are cared for vigorously. The (electron) balls he removes will be unexpectedly accelerated in every crash, and these acceleration charges propel electric waves. And, of course, metal electrons have long lines that are particularly long, the vibration of the shortcuts affects them the most, so they cannot function in a social event and transmit heat energy indefinitely. It is clear from such assumptions that large-scale radiation shields are the most acceptable manufacturers.

Indeed, we can be more accurate: the body emits rays at a given temperature and reappears just as it emits the same rays.

This has been demonstrated: the basic point is that if we think that a certain body can end up better than a transfer, then in a room full of things with the same temperature, it will add radiation from different bodies better than restoring energy to them. This means it will improve, and the rest of the room will be cold, rejecting the second law of thermodynamics. (We can use such a body to build a warm car that separates the filling as the room gets colder and colder!)

However, the metal shines when it is warm enough: why would it be so? As the temperature rises, the shortcut portion of the particles vibrates at a constant level; this movement disperses and speeds up electrons. Indeed, even glass is illuminated at temperatures high enough as electrons emit and move.

Two Basic Rules

The basic assumption is based on the radiation test view of the gap

Stefan's Law (1879)

Total P power from one square meter of black area in temperature T travels as a fourth total temperature:

$P = \sigma T_4, \sigma = 5.67 \times 10-8$ watts / sq.m. / K_4

Five years later, in 1884, Boltzmann discovered this T_4 behavior in theory: he used traditional thermodynamic thinking in a case loaded with electromagnetic radiation, using Maxwell's phenomena to associate the intensity and intensity of energy. (The minuscule ratio of the energy from the opening will obviously have a temperature-dependent similar to the radiation strength inside.) See what's going on in the notes on the subtlety of the decision.

Wien Relocation Act (1893)

As the temperature of the broiler shifts, so does the repetition, where the radiation is transmitted more frequently. That is also legally related to the overall temperature: $f_{max} \propto T$

(Wien himself discovered this law by speculation in 1893, after Boltzmann's speculation about thermodynamic. It had recently been observed, wherever it is, equally, by the American astronomer Langley.)

Truth be told, this high rise in fmax and T is natural for everyone — when the metal is heated in a fire, the main visible rays (about 900K) are reddish, with very little visible re-visible light. Further increase in T causes a darker shade from orange to yellow, eventually blue to higher temperatures (10,000K or higher) when high radiation exposure is clearly visible.

This is a repetitive step where the greatest force is important in maintaining sun-related energy, for example, in kindergarten. The glass should let the sun's rays in. However, not allow the heat rays to come out. This is understandable because the two rays are at a completely different frequency — 5700K and, state, 300K — and there are direct-to-light objects that are inaccurate in infrared radiation. Kindergartens operate because fmax changes with temperature.

Chapter 11
Entanglement

One of the basic and often seen as a strange occurrence in quantum mechanics is the concept of "entanglement." Entanglement is a phenomenon where two quantum particles interact appropriately, and their states will depend on each other, irrespective of how far apart they are.

Einstein's empirical study on momentum and position measurements led him to this concept. Indeed, entangled particles can appear remote and distant, yet they have the same physical structure, allowing interrelation and interconnection.

Another model to consider genuinely in our modern days is the interconnection and the polarization of the entrapped photons. It's clear that when the entanglement of electrons occurs, even though particles are not connected, they are dependent upon similar movements and happenings. Concretely, this model isn't completely perceived by physicists, even though it is considered a fundamental standard of quantum physics.

The spin may have a positive or negative (upward or downward) value (or direction, known as its "sign"), and when the electrons are entangled, it means that the measurement will show that their spin signs are opposite. If the spin of one of the entangled electrons is determined, it is immediately known that the spin of the second electron has the opposite sign. In reality, this entanglement occurs, for example, when the particles are formed in a single process (in the case of experiments with photons, identical, entangled photons are produced in the process of decomposition of one photon twice their size) or when they are components of one system, such as electrons being components of one atom.

In the case of entanglement, can these characteristics be coordinated with the speed of light rather than instantaneously? There are no methods to measure time and speed with absolute precision. However, instruments' precision is constantly improving, and modern experiments have shown so far that the speed of interaction between particles during entanglement exceeds the speed of light by at least 100,000 times! Scientists assume that if the speed of interaction during entanglement exceeds the speed of light (that much), then this interaction has infinite velocity, i.e., both particles acquire the exact characteristics simultaneously regardless of distance (non-locality).

For instance, if you measure two particles located in different countries and then measure simultaneously, the measurement carried out in one country or location will absolutely and undeniably decide what the outcome in the other location will turn out to be.

There is no theory to describe the correlation between these particles with certain states. It is known that the two states will remain indeterminate until one of the states is weighed. At this time, the states of each of the particles are then determined, not minding their distances apart. Many experiments have been carried out in the last thirty years using atoms and light to confirm this theory. The experiment carried out now still confirms the quantum prediction.

It is worthy to note that this does not, in any way, serve as a means to sending signals that are faster than light since the measurement in one location says Oxford will determine the state of the particle in another location, maybe Harvard. This shows that the outcome of each measurement is random. The particle at Harvard cannot be adjusted to match the result obtained for the particle at Princeton.

The connection between these two measurements can only be evident when the two sets of data are measured and compared, and this process has to be done at a speed that is quite slower than the speed of light.

Once a thing isn't forbidden, it is mandatory.

If a quantum particle moves from point A to point B, it will simultaneously take all the paths from point A to point B. This usually includes paths that have unlikely events like electron-positron pairs that appear from nowhere and disappear suddenly. A field of Quantum physics referred to as quantum electrodynamics (QED) implies that every possible process should be studied, including the very unlikely ones.

The QED is not just a random process of guessing without any real application.

The prediction of the interaction between the electron and the magnetic field by QED correctly describes the interaction at 14 decimal places. As strange as this theory may seem, it remains one of the best-tested hypotheses in the history of science.

Chapter 12
An Introduction to the Strings Theory Made Easy for Beginners

String theory is so named because it was created to describe the nature of quantum objects as something like a vibrating string, and those objects were measured based on those vibrations. Without getting too deeply into it, string theory involves a lot of work with subatomic particles, which are the smaller particles that makeup atoms (like quarks, for example). String theory was essentially formulated to help explain strange interactions that sometimes happened between some of the known subatomic particles. These particles would sometimes act like strings bound them together. Thus, string theory was formed.

Now, one of the most interesting parts of this early string theory was that these weird, string-based interactions worked out some mysterious math. For one, the vibrations in the strings predicted the existence of a certain particle: the graviton. It's the only quantum theory that has successfully done this so far. Theoretically, a graviton is a particle that causes gravity, and one of string theory's biggest advantages here is that its graviton is shaped like a donut. That may sound strange, but that same shape prevents many of the mathematical anomalies that arise when envisioning

An Introduction to the Strings Theory Made Easy for Beginners 53

gravity as a particle. However, we still can't prove that the graviton even exists with today's science. Gravity still presents many mysteries to us.

The string theory successfully described gravity as its quantum particle this way. However, string theory has many limitations that keep it from being an attractive (or logical) solution to the quantum universe's mysteries. For one, the final proposal of string theory requires ten dimensions to function properly, but we've so far only observed four dimensions in our existence. To remedy that, physicists and scientists alike have tried to make string theory work in its prerequisite ten dimensions, then remove the extra dimensions that don't apply to our universe. However, none have yet been successful. Besides, there are several other versions of string theory — there isn't just one. Bosonic string theory was one example, and it required twenty-six dimensions!

M-theory required eleven dimensions; so, while it managed to unify many of the past string theories into one neat package, it still didn't seem anywhere near reasonable science.

In addition, the "theory of everything," as its name implies, is an all-encompassing theory. It explains how quantum physics and quantum particles worked in one way or another. Although it's one of today's physicists and scientists' premier goals, we haven't formulated a working theory of everything yet. However, string theory makes a compelling case for becoming or at least supplementing a potential theory of everything.

Envision a guitar string in slow motion: the string moves in a wave shape, just like light or sound. The tension (how taut the string is pulled) of the string also controls what note any string on a guitar plays. If you pluck the string, it will vibrate. This is the prevailing theory behind the strings: each string has a different frequency (like the note of a guitar string) that it vibrates at, and each string also has a different length, which changes the number of notes that the string could play. Think about how, when playing the guitar, moving your fingers down the neck, shortening the string, will play higher notes while moving your hand up the neck will lengthen the string and play lower notes.

Interestingly enough, the intervals between these possible "notes" are also defined by a familiar variable: Planck's constant. Unfortunately, as convenient as this might seem, there's one question that scientists seem to be unwilling (or unable) to answer about string theory. What are the strings made of? Answers that have been tried are "pure mass" (whatever that means), irregularities in the fabric of reality, and the "energy" of "existence." However, the best answer is, "The answer is irrelevant because they're there."

One of the biggest downfalls of string theory is that, because it's so complicated and exists in so many additional dimensions (that we can't even find), we can't test it. However, we also can't rule it out entirely wrong. This creates a bit of a catch twenty-two: should scientists and physicists keep pouring person-hours and research into a theory that we technically can't even test, or should we give up on it and risk sidelining a potential theory of the universe? There's no good answer to this question, but it's one of the pitfalls that plague string theory and those interested in it.

Despite its shortcomings, though, many scientists still believe that string theory is the way things are, or, at least, it's a step to figuring it out. However, until we find six or seven more dimensions, we'll have to wait on that.

Chapter 13
Black Holes and Spacetime

Black Holes and the Mystery of Quantum Gravity

Einstein's theory is technically irreconcilable with quantum physics for several reasons that we've mentioned in this book. According to Einstein, black holes are regions of ultra-intense gravity within space-time that are so powerful — they pull even light inside.

The idea of a black hole was first proposed by a clergyman named John Mitchell in 1784. His ideas were dismissed for the most part because, shortly afterward, the light was "discovered" to be a wave instead of a particle (at the time). Therefore, scientists were unsure if gravity would act on light "waves" instead of "particles" and thus largely forgot about the concept. That is until Einstein brought them back in 1915 with his general theory of relativity. From there, many physicists have hotly debated exactly how black holes are formed, how they work, and how they influence space and time. Scientists believe that time is so distorted in a black hole that, to an outside observer, time would appear to stop inside it. However, if a person were to fall into a black hole, time would proceed for them normally (resulting in a rather gruesome death).

Black holes can be created in a few different ways. One way would be by violating Planck's law. This has to do with the Planck constant, which you learned about before. Planck's constant also defines several measurements within quantum physics, and these measurements denote certain limits of measurement that can be achieved within the quantum universe. For example, if you tried to measure a particle's position with a laser at a greater accuracy than one Planck length (1.6×10^{-35} meters, which is very, *very* small), the laser's power would end up creating a very small black hole. Ironically, the black hole created would be exactly the size of one Planck length. Since time and space are intertwined, a black hole can also be created when measuring a length of time less than one unit of Planck time (10^{-43} seconds, which is very short).

More traditionally, black holes are created by the collapse of very large stars. At the end of a star's life cycle, when the fuel inside the star has all run out, the mass of the star itself causes it to collapse, and all the matter inside of the star gets sucked in. Sometimes, this can create smaller, dimmer stars or create quasars (a type of ultra-bright, fast spinning, and very small star containing a black hole). Remember Heisenberg's uncertainty principle? Theoretically, this also can apply to the creation of black holes. If you'll remember, according to the uncertainty principle, it's impossible to determine a quantum particle's momentum and position with high precision at the same time.

If you attempt to measure one or the other more precisely, the other measurement becomes less and less precise. If you get precise enough — say, down to one Planck length or less — the other value becomes so large that the particle is mathematically capable of turning into a black hole. This doesn't mean the particle is becoming a black hole, but more of a math problem illustrating quantum particles' difficulty. This is called an absurdity or a nonsense result; according to physicists, this means that something is missing from the math. They're missing a formula, a variable, or something else. This is the mystery of quantum gravity — even today, we're still missing something in the math required to make gravity (and, by extension, black holes) make sense.

These issues are propagated because there is no way to measure gravity on a coordinate plane. Since gravity exists within four dimensions, it can't be figured out using the two-dimensional math that scientists have traditionally used to figure out other quantum theories. Additionally, if you try to apply four-dimensional gravity to these two-dimensional theories, you receive anomalous answers—the math just doesn't make sense. Essentially, gravity is space-time. As a result, you can't anchor any math within space-time to help figure it out.

The general theory of relativity works at a level that we can see, but it unravels at the quantum level (i.e., below the Planck scale). This brings us full circle because that's why quantum mechanics was created in the first place. However, quantum mechanics has yet to account for the anomaly of gravity. For now, physicists continue to assume that we simply haven't found the right theory to solve it yet.

To a lesser extent, black holes (and everything in the universe that experiences gravity) experience gravitational time dilation. This means that the closer you are to the black hole center, the slower time moves to an observer outside the black hole. However, strangely enough, time will proceed normally to those closest to the black hole. This means that, if a pair of twins were to stand by a black hole — one twin very close to the black hole, and the other very far away — when the twins both came back to earth, the twin who was further from the black hole will have aged more than the one who was close to the black hole. Crazy, right? To a lesser extent, this also happens to the satellites orbiting the earth and the international space station (the ISS). Clocks that orbit Earth from space will slowly move ahead of clocks on earth.

Even stranger — continuing with the twin metaphor — if one of the twins were to fall inside the black hole, and the other twin was to watch, time would appear to completely stop on the twin inside the black hole as the twin crossed over the edge. The twin watching outside the black hole would just see their other twin stop, freeze totally in time, then never move again, no matter how long the other twin waited for them. No one knows whether it would be possible to escape a black hole after this had happened if it were possible to survive the experience.

According to all known laws of physics, any normal person going into a black hole would be crushed by the incredibly strong gravitational force long before they made it to the center — of course. One theory is the white hole theory, which says that for every black hole, there is a matching white hole somewhere else in the galaxy that spews out all of the matter consumed by the black hole. According to this theory, if the person who fell into the black hole survived, then the black hole would work as a teleportation device. However, although we have documented cases of proposed white holes in the universe (the white hole GRB 060614 was found in 2006), there's no way to prove that these holes are in any way connected to black holes.

Some scientists have proposed that the Big Bang that created the universe was a white hole. The same paper proposes that white holes should be spontaneous, limited occurrences, rather than long-lasting singularities like black holes.

Chapter 14
Most Popular Experiments Explained

Unified Energy and Unified Matter

This generates us to the concept of the Unified Field. This also brings us to the Unified Force, Unified Matter, and the Theory of Everything.

In this part, I offer that the "energy string" is the fundamental entity of the universe. These energy strings are the special hybrid of mass and energy from which all the types of particles and forms of energy are created.

I also believe that the five types of energy strings provide most, if not all, of the properties found in the universe.

Therefore, this leads me to believe the above set of concepts will ultimately move us toward the "Unified Field Theory."

Unified Field

In molecular materials science, the unified field hypothesis is a push to characterize each primary power and the connections concerning elementary particles in the specifications of a singular hypothetical structure. In material science, powers can be named by fields that referee dealings among disengaged things. During the nineteenth century, James Clerk Maxwell passed on the underlying field hypothesis in his way of thinking of electromagnetism. At that point, in the early part of the twentieth century, Albert Einstein set up general relativity, the field hypothesis of attractive energy. Consequently, Einstein and others attempted to create a bound together field hypothesis in which electromagnetism and gravity would surface as various parts of a solitary major field. They fizzled, and right up 'til today, gravity stays past endeavors at a bound together field hypothesis.

At subatomic separations, fields are portrayed by quantum field hypotheses, which apply the thoughts of quantum mechanics to the whole ground. During the 1940s, quantum electrodynamics (QED), the quantum field hypothesis of electromagnetism, grew settled entirely. In QED, charged particles associate as they produce and retain photons (minute parcels of electromagnetic radiation). As a result, trading the photons in a round of subatomic "get." This hypothesis works so well that it has become the model for speculations of different powers.

Moreover, I have said that the "field" is just the streaming of vitality strings. Accordingly, on the off chance that all "fields" are kinds of vitality strings, at that point, we can make a "Brought together Field Theory" in light of "bringing together" these vitality strings.

In particular, I accept that these sorts of vitality strings are varieties of one essential kind of vitality string. I see the kinds of vitality strings recorded and portrayed above as branches off a tree. We know we have comparable branches. We should simply locate the principal part of the tree.

When we distinguish that focal part, we will have the option to bring together all fields into one complete picture.

Unified Matter

These energy strings also lead us to Unified Matter. Many scientists, including Heisenberg, talk not only of a unified field but a "Unified Matter." Heisenberg believed that the quarks being discovered actually had similar properties and, therefore, would lead us to understand a Unified Matter.

Indeed, this we can do with our energy strings. The energy strings I have proposed are the hybrid of energy and matter. Therefore, by unifying these energy strings, we will not only be unifying our energy fields but also unifying our matter.

The Future of Unified Field-Based on Energy Strings

I do believe that the future of the Unified Field, the Unified Force, and the Unified Matter will all be developed from the five fundamental types of Energy Strings I have listed and mentioned above.

For the first time, we have a common language and a common structure for each type of field, each type of force, and all the observed properties in any particle. For the first time, we have commonalities, where every field, force, and particle can begin to look like the other fields, forces, and particles. This is a big step.

Now we can take these commonalities and find the True One Unifying Concept behind them all. Indeed, from just One Universal Energy, we can create all things. All energies, matter, motions, and observations can be explained based on these few concepts.

These secrets have now been discovered. They will gradually be revealed through a future series of publications, leading ultimately to the Unified Energy Solution and the Theory of Everything.

Superconductors

No Resistance

When does resistance disappear? The answer to this question was made by Kamerlingh Onnes as early as 1914. He proposed a very ingenious method of measuring resistance. The experimental scheme looks quite simple. Lead wires from the coil in a cryostat omitted — an apparatus for conducting experiments at low temperatures. The cooled helium coil is superconducting. In this case, the present flows through the coil, creating a magnetic field around it, which can be easily detected by the deviation of the magnetic needle located outside the cryostat. Then close the key so that now the superconducting stroke was short-circuited. However, a compass needle was diverted, indicating the presence of current in the coil is already disconnected from a current source. Watching the arrow for a few hours (until evaporating the helium from the vessel), Onnes had not noticed the slightest change in the deflection direction.

According to the experiment results, Onnes concluded that the resistance of the superconducting lead wire of at least 10^{11} times lower than its resistance in the normal state. Subsequently, conducting similar experiments, it was found that the current decay time exceeds many years, and this indicated that the superconductor resistivity is less than 10^{25} ohms · m. Comparing this with the resistivity of copper at room temperature, 1.55×10^{-8} ohm-m, the difference is so large that one can safely assume that: the resistance of the superconductor is zero, it is difficult to call another monitor and modify the physical quantity which would be addressed in the same "ground zero" as the conductor resistance below the critical temperature.

Recall known from school physics course Joule — Lenz: the current I flowing through the conductor with a resistance R it generates heat. At this consumed power $P = I^2 R$. As little resistance to metals, it often limits the technical possibilities of different devices. Heated wires, cables, machines, apparatus, therefore, millions of kilowatts of electricity thrown to the wind. Heating limits the throughput power, the power of electric cars. Thus, in particular, is the case with electromagnets. Obtaining strong magnetic fields requires a large current, which leads to the release of an enormous amount of heat in the windings of the solenoid. But the superconducting circuit remains cold, and the current will circulate without damping — impedance equal to zero, no power losses.

Since the electric resistance is zero, the exciting current of a superconducting ring will exist indefinitely. The electric current, in this case, resembles the current produced by the electron orbit in the Bohr atom: it is like a very large Bohr orbit. Persistent current and the magnetic field generated by it cannot have an arbitrary value; they are quantized so that the magnetic flux penetrating the ring takes values that are multiples of the elementary flux quantum F on = h / (2e) = 2.07 10 15 Wb (h - Planck's constant).

Unlike electrons in atoms and other micro-particles whose behavior is described by quantum theory, superconductivity — macroscopic quantum phenomenon. Indeed, the length of the superconducting wire through which flows persistent current can reach many meters or even kilometers. Thus, carriers describe a single wave function. This is not the only macroscopic quantum phenomenon. Another example is superfluid liquid helium or substance neutron stars.

Electrical Resistance Superconductors

No experimental methods are fundamentally impossible to prove that any value, particularly the electrical resistance, is zero. It can only show that it is less than a certain value determined by the measurement accuracy.

The most accurate method of measuring low impedances consists in measuring the current decay time induced in the closed circuit of the test material. A decrease in current time energy LI 2 /2 (L - inductance loop rate) consumed for Joule heat:

Here integrating (I o - current value at t = o, R - resistance of the circuit).

The current decays exponentially with time, and the electrical resistance determines the attenuation rate (at a given L).

Here dI- current change during Δt. Experiments conducted using a thin-walled superconducting cylinder with extremely small values of L showed that the superconducting current is constant (with accuracy) within a few years. It followed that the resistivity in the superconducting state is less than 4 · 10 25-ohm m or more than 10 17 times less than the resistance of copper at room temperature. Since the possible decay time is comparable to the time of the existence of humanity, we can assume that R DC in the superconducting state is zero.

Thus, the superconducting current is only like a real-life example of perpetual motion on the macroscopic scale!

When R = 0, the potential difference V = IR on any segment of the superconductor, and hence the electric field E inside the superconductor is zero, the electrons create a current in the superconductor, moving at a constant speed without being scattered by the thermal vibrations of the crystal lattice atoms and their irregularities. Note that if E is not equal to zero, electrons carrying the superconducting current accelerate without limit, and the current could reach an infinitely large value, which is physically impossible.

The situation changes if the superconductor is applied to the variable potential difference to create a variable superconducting current. During each period, the current changes direction. Consequently, the superconductor must exist in an electric field, which periodically slows down the superconducting electrons and accelerates them in the opposite direction. Since it consumes energy from an external source, the electrical resistance of alternating current in a superconducting state is zero. However, because the electron mass is very small, power loss at frequencies less than $10^{10} - 10^{11}$ Hz is negligible.

Tunnel Effects

In 1962, an article appeared before anyone unknown author B. Josephson, who theoretically predicted the existence of two extraordinary effects: steady and unsteady. Josephson theoretically studied the tunneling of Cooper pairs from one superconductor to another through any barrier. Before proceeding to the first Josephson Effect, briefly the tunneling of electrons between the two metal parts separated by a thin dielectric layer.

The tunnel effect has been known in physics for a long time. The tunnel effect — this is a typical problem of quantum mechanics. Particle (for example, an electron in the metal) approaches the barrier (e.g., a dielectric layer) to overcome which her classical ideas cannot, as its kinetic energy is insufficient. However, the barrier area with his kinetic energy could well exist. On the contrary, the barrier passage is possible according to quantum mechanics. The particle may have a chance, as it were, to pass through the tunnel through a classically forbidden region where its potential energy would be more like a full, i.e., the classical kinetic energy as it is negative. Quantum mechanics for the microparticles (electron) holds uncertainty relation $\Delta h \Delta r > h$ (x - coordinate of the particle, p - its pulse). When a small uncertainty of its coordinates in a dielectric $\Delta h = d$ (d - thickness of the dielectric layer) leads to large uncertainty, its pulse $Dp \geq h / \Delta x$, and consequently, the kinetic energy $p \, 2 / (2m)$ (m - a mass of particles), the energy conservation law is not violated. Experience shows that indeed between two metal electrodes separated by a thin insulating layer (tunnel barrier), electric current can flow the greater, the thinner the dielectric layer.

Josephson Effect

Physical objects in which the Josephson Effect takes place, now called Josephson junctions or Josephson junctions, or Josephson elements. To imagine the role played by Josephson elements in superconducting electronics, it is possible to draw a parallel between them and the semiconductor p-n-junctions (diodes, transistors) — element base conventional semiconductor electronics.

Josephson junctions are some weak electrical communication between two superconductors. This linking can be made in several ways. The most commonly used types in practice weak link is: 1) tunnel junctions, in which the bond between the two film superconductors is carried out through a very thin (tens of angstroms) insulating layer between them — SIS-structure; 2) "sandwiches," two film superconductors interacting through a thin (hundreds Angstroms) layer of a normal metal therebetween — SNS-structure; 3) the structure of the bridge type, which is a narrow superconducting bridge (bridge) of limited length between two massive superconducting electrodes.

Bearers overcurrent superconductors at $T = 0$ K are all conduction electrons n (o) (electron density). When the temperature rises, elementary excitation (normal electrons) appears so that the concentration of n s of the superconducting electrons at a temperature T.

n s (T) = f (o) -n n (T),

where n n (T) — the concentration of electrons at a normal temperature T. The Bardeen-Cooper-Schrieffer (BCS) for T → T c (critical temperature).

p s (T) ≈ Δ 2 (T)

where 2 D (T) — the width of the energy gap in the spectrum of the superconductor. All superconducting electrons form pairs associated state, known as Cooper pairs of electrons.

Cooper pair combines two electrons with opposite spins and pulses and has a zero net spin. Unlike normal electrons having a spin of 1/2 and therefore obey Fermi-Dirac statistics, Cooper pairs obey Bose-Einstein statistics and condensed at one, the lower energy level. A characteristic feature of Cooper pairs is their relatively large size (about 1 micron) is much greater than the average distance between pairs (of the order of the interatomic distances). Such strong spatial overlap pairs mean that all of the (condensation) of the Cooper pairs is coherent, which is described in quantum mechanics, the wave function of a single W =. DELTA.E ix. Here A - amplitude of the wave function, the square of which characterizes the concentration of Cooper pairs, h - the phase of the wave function, i - imaginary unit, P - -1. In the case of normal electrons, which are fermions, the Pauli Exclusion Principle, the electron energy is never exactly equal to each other. Therefore, the Schrodinger equation for these particles follows that the phase velocity dq / dt of the wave functions of electrons normal differ; thus, phase h are uniformly distributed in the trigonometric circle, and the summation over all particles explicit dependence on h disappears.

The presence of a weak electric connection between the superconducting electrodes due to poor overlap of the wave functions of the Cooper pairs of electrodes, whereby such contact is also superconducting, but the density of the critical current value is much (by several orders of magnitude) smaller than the critical current density of the electrodes $j_c \approx 10^8$ A/cm^2. For tunneling structures and structures of Josephson's sandwich-type critical current density, junctions-ing is typically in diapason j_c from 10^1 to 10^4 A/cm^2, and their area S_B within modern technology can be made from a few hundred to a few square microns. Therefore, the critical current of the Josephson element $I_c = j_{jc}.S$ may be from a few milliamperes to a few microamperes.

Chapter 15
Quantum Computing

What is a Computer?

A computer is a machine that receives and stores information input, processes the information according to a programmable sequence of steps, and produces the resulting output of information. The term 'computer' was used for the first time in the 1600s to refer to people who perform calculations or computations and now refers to computers that compute. Computing machines can be divided roughly into four types:

1. Computing devices for classical computational physics. These machines use moving parts, including levers and gears, to perform computing. Usually, they are not programmable but always perform the same operation, such as adding numbers. An example is the 1905 Burroughs Adding Machine.

2. Electromechanical classical mechanics fully programmable computing devices. These machines operate using electronically controlled moving parts. They process information stored as digital bits represented by the locations of many electromechanical switches. The first such machine was built in 1941 by Konrad Zuse in

wartime Germany. In theory, their programmability allows them to solve any problem that can be found and overcome by using algebra. These were the first 'universal' computers in this context.

3. All-electronic, hybrid quantum-classical-physics computers. These fully programmable, universal computing machines have no moving mechanical parts and work using electronic circuits. The first to be constructed was the ENIAC, engineered by John Mauchly and J. Presper Eckert, University of Pennsylvania, 1946. The physical principles that describe the motion of electrons in these circuits are rooted in quantum physics. But, given that there are no superposition states or entangled states involving electrons in different circuit components (capacitors, transistors, etc.). Therefore, we call these machines — essentially any computer in operation today — 'classic computers.'

4. Quantum computers. If ever built successfully, these devices would operate according to the principles of quantum physics. Knowledge will be expressed by the quantum states of individual electrons or other elementary quantum artifacts, and there will be entangled states involving electrons in various circuit components. These computers are expected to solve those kinds of problems much faster than any modern classical computer can do.

How do Computers Work?

Computers store and manipulate information using a binary alphabet language consisting of only two symbols: 0 and 1. Every 1 or 0 symbol is referred to as a bit, short for a binary digit because it can take two possible values. A page of text, such as the one you read, is represented as a long string of numbers in a computer file. A binary code represents every letter. For example, 'A' becomes 01000001, 'B' becomes 01000001, and so on.

In a typical computer, each bit is represented by the number of electrons stored in a small device called a capacitor. We can think of a capacitor as a box that holds a certain number of electrons, sort of like a bulk grain bin in a food store that holds a certain amount of rice. Each capacitor is called a memory cell. For example, such a capacitor could have a maximum capacity of 1,000 electrons. If the capacitor is full or almost full of electrons, we say it represents a bit of a value of 1. If the capacitor is empty or almost empty, it represents a bit value of 0. The capacitor is not allowed to be half-filled, and the circuitry is designed to ensure that this does not happen. By grouping together eight capacitors, each of which is either full or empty, any eight-bit number — e.g., 01110011 — can be interpreted.

The role of the machine circuitry is to empty or fill various capacitors according to a set of rules called a program. Eventually, the action of filling and emptying the capacitors manages to perform the desired calculation — say, to add two 8-bit numbers. In a computer, tiny components of computer circuitry called logic gates perform the actions of a computer. A logic gate is made of silicon, and other elements arranged to either block or pass electrical charge, depending on its electrical environment. Logic gate inputs are bit values, represented by a full capacitor (a 1) or an empty capacitor (a 0). (The word 'gate' is associated with the fact that something goes into it and something comes out of it.)

How small can a single logic gate be?

In the first all-electronic computers, such as the ENIAC, built in the 1940s, a single logic gate was a vacuum tube similar to the amplifier tubes still used today in vintage-style electric guitar amplifiers. Each tube has at least the size of your thumb. By 1970, the microcircuit revolution was able to reduce the size of each gate to about one-hundredth of a millimeter. When things get much smaller than this, it's best to measure the length of a unit called a nanometer, which is equal to one-millionth of a millimeter. The size of the gate in 1970 was 10,000 nanometers. On the other hand, a single silicon atom, which is the main atomic element in computer circuitry, is around 0.2 nanometer in thickness. By 2012, single gates in typical computers had been reduced enough to be spaced apart by as little as 22 nanometers — that is, only about a hundred atoms apart. The actual working area of the gate was less than 2.2 nanometers or 10 atoms in thickness. This small size allows you to place a few billion memory locations and gates in an area the size of your thumbnail.

Having gate sizes much smaller than those dimensions leads to both a curse and a blessing. We leave the domain of many-atomic physics and enter the domain of single-atomic physics. There are now variations between the classical physics principles that explain the average behavior of many atoms and the quantum physics principles required when dealing with single atoms. We reach a random action domain that doesn't sound good if we try to get a well-regulated system to do our numerical bidding. In reality, a group of scientists led by Michelle Simmons, director of the Center for Quantum Computation and Communication at the University of New South Wales, Australia, constructed a gate consisting of a single phosphorus atom embedded in a silicon crystal tube. This is the smallest gate ever to be designed. This gate only functions properly if cooled to an extremely low temperature: – 459 degrees Fahrenheit (– 273 degrees Celsius). Suppose the material is not at least as cold. In that case, the random (thermal) motion of the silicon atoms in the crystal decreases the confinement of the electron psi wave, which may leak out of the channel into which it is intended to be confined. For day-to-day desktop computers, which, after all, have to operate at room temperature, this leakage prevents such single-atom gates from being the basis of technology that everyone can use. On the other hand, these experiments demonstrate that at least in theory, computers can be built on the atomic scale, where quantum physics rules.

Can we create computers that fundamentally use quantum behavior?

Given that physics defines the ultimate behavior and efficiency of information transfer, storage, and processing, it is reasonable to ask how quantum physics plays a role in information technology. Because electronic computers rely on the behavior of electrons, and communication systems rely on the behavior of photons — both elementary particles — it is not surprising that quantum physics ultimately determines the performance of information technology. But here is subtlety. Current computer technologies do not involve quantum superposition states to represent information. They use states considered classical states of physical things — namely, groups of electrons.

The big question is: can we create computers that use quantum-mechanical states to enhance our ability to solve real-world problems? If these computers were ever built, they would be able to bypass certain forms of data encryption methods much faster than any computer operating today. This would revolutionize the field of privacy and confidentiality for computers and the Internet. The encryption key that might take thousands of years to crack using a conventional computer could only take minutes on a quantum computer.

What is a Qubit?

The word bit refers to both the abstract, disembodied mathematical concept of information and the physical entity that embodies the information. It is clear in classical physics that a 'physical bit' carries one 'abstract bit' of knowledge. There is a very simple one-to-one relationship between the state of the physical bit and the value of the abstract bit, 0 or 1. We may also use individual quantum artifacts, such as an electron or a photon, to incarnate a portion. In this case, the elementary physical entity is called a qubit, short for 'quantum bit.' A qubit has two different quantum states, such as the H and V polarization of the photon, or the upper path and the lower path of the electron. When measured, the results represent a bit value of 0 or 1. But remember that we can select different polarization measurement schemes — say, H/V or D/A.

The results may then be random, with the probability of observing possible outcomes, depending on which measurement scheme we selected. In this case, there is no one-to-one relation between the state of the physical qubit and the value of some abstract conceptual bit. The concepts of quantum physics suggest significant variations between the behavior of classical bits and qubits. Classical bits can be copied as many times as we want, without any degradation of the information; qubits cannot be copied or cloned even once, although they can be teleported. The state of the classical bit, 0 or 1, can be determined by a single measurement; any sequence of measurements cannot determine the quantum state of the single qubit.

What physical principles differentiate classical and quantum computers apart?

There are large differences between the types of gates used in classical computers and the gates that need to be used in quantum computers. Classical gates perform operations that are not reversible; understanding the output does not tell you what the inputs are. On the other hand, if a quantum gate is to operate properly with qubits, it must be reversible. That is, you need to be able to determine the input states by understanding the output states. This requirement arises because any quantum gate operation must be a unit process. Recall that we use the word 'unitary' to refer to physical processes or behaviors that cannot be divided into individual steps, each with definite, observable outcomes. Our main example was an electron (or photon) passing from source to final location in a situation where two separate paths are possible. We pointed out that if there is no permanent trace left by the passage of the electron that indicates that it has taken a particular identifiable path, it is wrong to say that it has actually taken one path or the other. It's also not correct to say that both directions have been taken. The whole process of leaving one place and arriving at another must be seen as an undivided, complete process — that is, a unitary operation. These processes are reversible.

What logic gates would quantum computers use?

Since the early 1990s, scientists have been theorizing how a universally programmable computer based on quantum superposition and entanglement could be constructed. For which kinds of problems would it be ideally suited. Neither is it a simple issue, nor has it been completely resolved to date. On the other hand, considerable progress has already been made. The chances seem reasonable to decent that such a machine will become a reality within, say, ten or twenty years. A quantum computer takes qubits as inputs, performs a series of gate operations on them according to a program designed by the programmer, and outputs the modified qubits. The number of output qubits must be equal to the number of input qubits for the whole process to be unitary. As is the case with classical computers, there are different choices that the designer will make for the set of gates to be used in a quantum computer. I am focusing here on a set that is similar to the logic of {XOR, AND} described above for classical computers. For quantum computers, I follow the logic I call {QXOR, QR}. Using two quantum gates, called 'quantum XOR' and 'quantum ROTATE,' a universal quantum computer can be built, at least in principle. What are these two kinds of gates doing?

First, note that we give the two possible qubit states the names o and 1, and they are represented, for example, by the H-polarized and V-polarized states of a single photon. The general operating principles of a quantum computer are independent of how we choose to represent the qubits of physical objects.

The quantum XOR gate or the QXOR gate is shown in FIGURE 2.1. In the QXOR gate, the B qubit moves unchanged through the gate (from left to right) rather than being discarded, as in the classic example. This makes the QXOR gate both reversible and unitary. That is, outputs are uniquely linked to inputs.

A good way to think about the QXOR gate is to say that qubit B controls what happens to qubit A, as indicated by the arrow pointing from B to A. If qubit B is in the state (o), then qubit A passes through unchanged, as in parts I and (ii) of FIGURE 2.1. But if B is in the state (1), then the state of qubit A is changed from (o) to (1) or from (1) to (o) as in FIGURE 2.1 sections (iii) and (iv).

Figure 2.1 I Quantum XOR gate. Qubit B controls the process of qubit A modification. The Qubits are moving from left to right.

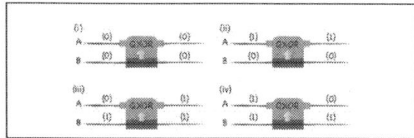

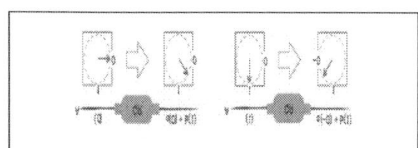

The second type of logic gate required is the quantum ROTATE gate, or QR gate, shown in FIGURE 2.3. This gate has one input and one output and is both reversible and unitary. If the input state is (0), the output state is the same superposition of the (0) and (1) states, with the state arrow pointing in the diagonal direction. If the input state is (1), the output state is again the same superposition of the (0) and (1) states, but with the state arrow pointing in the antidiagonal direction. Both 'a' and 'b' arrow components have a value of 0.707 (i.e., one-half square root). According to Born's Law, this implies the likelihood of obtaining either the outcome, (0) or (1), of measuring each equal to 0.5 or 50%.

Figure 2.3. The quantum-ROTATE gate produces (0) and (1) qubit states superposition. The arrow-rotation diagrams are similar to the polarization state arrow diagrams.

These two potential output states are similar to diagonal and anti-diagonal polarization. If a photon polarization represents the qubits, the QR gate is easily implemented using a special crystal that rotates the polarization state arrow by forty-five degrees in the counter-clockwise direction. Quantum computing theorists have shown that any qubit-based computation can be done by combining sequences of QXOR gates and QR gates.

How Would Quantum Computers Operate?

The classical machine works by defining input data in the form of bit selection and sending those bits to the processor where the gates function sequentially according to the program and then reading the bit values at the output.

The quantum computer works by defining input data in the form of a set of qubits, each with its quantum state specified, sending those qubits to the quantum processor where the gates act on them according to the program and then measuring the qubits at the output. The main difference between the classical case and the quantum case is that there can only be overlap and entanglement in the quantum case. These quantum states can only occur in the circuit if the overall operation of the gates together is a unitary process. The process is unitary only if there is no way, even in principle, for a person to know any of the individual qubit values (0 or 1) in the inner part of the circuit.

Which physics and chemistry problems can quantum computers solve?

The history of quantum computing did not begin with computer science but with physics. In 1981, Richard Feynman, one of the most inventive theoretical physicists, found out that the fundamental equation of quantum theory — Schrödinger's equation — cannot be solved efficiently by ordinary computers. Schrödinger's equation plays a role in quantum theory like Newton's motion laws in classical physics theory. The difference is that while Newton's laws explain how classic objects behave in terms of definite and perfectly predictable outcomes, Schrödinger's equation describes how quantum states shift in time.

Again, note that quantum states are not in one-to-one correspondence with the measurement results but show only the potential for results. The fact that Schrödinger's equation cannot be easily overcome by using ordinary computers is a big obstacle to the advancement of science. We have a fundamental equation that we need to solve to predict the probabilities of experimental outcomes; but, in the case of sufficiently complicated situations involving many quantum objects, we can't solve it! We simply don't know exactly what the theory predicts, so we can't make full use of it to advance science, engineering, and medical research. We can't develop better quantum-based drugs because it's impossible to solve Schrödinger's equation for large molecules.

Of course, scientists have many ways of finding approximate solutions to Schrödinger's equation, which is very helpful, but we don't have accurate solutions that might contain welcome surprises. Feynman assumed that a new type of computer he called a quantum computer would be able to solve Schrödinger's equation efficiently. Since Feynman pointed this out, a lot of work has gone into building a computer like this. Such a computer would operate according to quantum principles rather than the classical physics principles, as ordinary computers do. Unlike most computer science problems and math problems, problems involving Schrödinger's equation can easily be turned into algorithms that can be performed on a quantum computer. That's because Schrödinger's equation is the fundamental equation of quantum theory!

For example, Schrödinger's equation allows us to calculate the energy and shape of the psi wave for each possible quantum state of the electron within the atom. Molecules, such as the all-important DNA molecule, are made of atoms arranged in ways that create their structure and enable them to perform their functions, such as encoding and propagating a person's genetics. Because DNA molecules contain so many quantum particles — electrons, protons, and neutrons — using classical computers, it is impossible to precisely understand and predict the structure and function of Schrödinger's equation.

Consider a simple example to see why Schrödinger's equation for a DNA molecule using a classical computer is so difficult to solve. Let's assume that the molecule contains a minimum of five hundred electrons. This is a relatively small molecule compared to DNA. To represent the quantum state of these five hundred electrons using the bit state of the computer, it is necessary to represent all possible entangled states of the five hundred electrons.

Each of these possible states represents a quantum possibility for a particular combination of results that could be observed if measurements were made on all electrons. To keep it simple, let's say that each electron could be in one of two states, labeled 0 or 1 — that is, it can be thought of as a qubit. There are four possible variations if there are two electrons: 00, 01, 10, and 11. There are eight potential variations if there are three electrons: 000, 001, 010, 011, 100, 101, 110, and 111. There are 2^{500} or about 10^{150} possible combinations of states for five hundred electrons that need to be considered. This is much greater than the total number of elementary particles in the whole universe! Each combination must be represented by a number in the computer's memory, but it is impossible to store all of these numbers in any computer smaller than the entire universe. A solution could be to split all the combinations into smaller groups and process each group individually by moving the numbers into and out of the computer's memory. But the time needed to make all this move would probably take longer than the life of the universe. This example illustrates Feynman's main point: as the size of the quantum problem to be solved larger, the size of the computer needed to solve it grows even faster — exponentially.

Chapter 16
Ways Quantum Physics Affects Your Daily Life

Quantum technologies have increased, and today we cannot drive to grandma's house or buy food without taking advantage of quantum physics. Nevertheless, today's quantum technologies pale compared to potential new avenues in the future. The applications of quantum physics to improve health, faster computers, and safer communications are behind the horizon.

The Neon Light

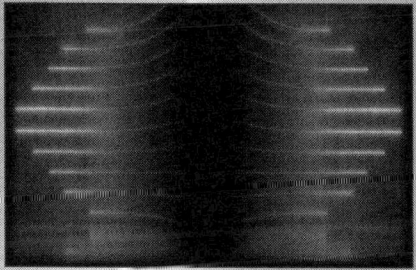

Neon light was first demonstrated in 1855 by the German physicist Heinrich Geissler. He noted that a slight glow was emitted when an electric field was applied through a gas tube containing low-pressure gas.

Nowadays, we know that the applied electric field was stripping the electrons from the atoms in the gas and creating a flow of negative electrons in one direction and positively charged ionized (ions) atoms in the other.

Collisions between fast-flowing ions or electrons with atoms lead to other ionizations, thus continuing the process. This set of electrons and ions is called plasma. When we think of solids, liquids, and gases as the only three states of matter, physicists consider plasma as a fourth state.

Collisions between ions or electrons and atoms do not always have enough energy to release atomic electrons. Colliding atoms are only empowered from their ground state to an excited state when they do not. Shortly after that, the atom will transition down to a lower-lying state, thus emitting a photon at a frequency set by the spacing between the energy levels.

These photons are responsible for the glow, and the characteristic frequency defines its color. Red light is emitted when a light discharge uses neon, while helium emits a purple color, carbon dioxide emits a white color, and mercury emits a blue color.

This physical process inspired French engineer Georges Claude to formulate a patent for the technology in France in 1910, which led to the use of neon lights for advertising and art. The light discharge is also the basis of the sodium vapor lamp, whose light yellow-orange glow is used to illuminate many streets around the world.

A miniaturized device that operates on the same principles, the neon incandescent lamp, was introduced in 1917. These were used in the 1970s for electronic displays and today serve as the necessary technology for televisions and plasma displays.

Even ordinary everyday fluorescent bulbs are based on light discharge. Here, the exhaust emissions come from mercury vapor, which are in the (invisible) ultraviolet range.

The Laser

The laser is one of the best examples of quantum application because it is widely used.

We have already seen that excited atoms emit photons by making a quantum leap to a lower energy state. In most cases, this occurs without any external influence, and the emissions of this variety are called spontaneous.

This is only half the story since atoms can also be driven to emit photons with a process known as stimulated emission.

Stimulated emission is an essential physical process first envisaged by Einstein in 1917.

Before it became so used, "laser" was an acronym for "light amplification through stimulated emission of radiation."

The American physicist Theodore Maiman developed the first successful laser in 1958. He found a similar effect, "microwave amplification by stimulated radiation emission." The short name given to this was the maser, and its realization was awarded the Nobel Prize in 1964.

The stimulated emission of a photon occurs when a photon of the same frequency hits an excited atom; that is, a photon stimulates the emission of an equal photon.

Furthermore, the stimulated photon has the same frequency as the incident photon and emerges in the same direction and phase.

The stimulated emission requires an incoming photon and an

excited atom. The excited state's energy must be $\Delta E = hf$ above the ground state (h is Planck's constant, f-frequency, and E-energy).

Otherwise, the atom cannot make a quantum leap and emit a photon at the same frequency.

Now imagine that you have an extensive collection of atoms and that many of them are excited $\Delta E = hf$ above the earth state. Introducing a photon at frequency f can stimulate a quantum jump in one of the excited atoms to end up with two equal photons. These two photons can then go on and stimulate two other identical photons, resulting in four "cloned" photons. If you have a vast population of excited atoms, you will have a large army of equal and cloned photons due to the waterfall effect.

This is precisely the "amplification" of the stimulated emission from which the laser derives its name.

Our description of amplification is based on having many excited atoms.

However, atoms prefer to be in their ground state. So, something needs to be done to prepare an extensive collection of excited-state atoms. This is called population reversal, and there are many ways to achieve it. However, it is necessary to have more atoms in the excited state than in the earth state in all cases. Otherwise, more photons will be absorbed than those emitted, and the waterfall will run out.

The process can be helped by placing a pair of mirrors on each end of the collection of atoms. One mirror should be 100% reflective, while the other should be partially transparent. This will allow a coherent, one-way beam of light to escape from one end and continue doing useful things. The first laser in the world consisted of a flash lamp coiled around a ruby rod inserted between a pair of mirrors. The flashing lamp generated a population reversal of atoms in the ruby crystal, and mirrors bounced photons back and forth to build a cascade of stimulated photons.

Today, lasers are available in all shapes and sizes.

The population of atoms, or "laser medium," is usually some form of gas or solid. Also, different lasers use different population inversion schemes. Lasers can also be set to emit light continuously or in pulses, and the pulse's energy and the pulse's duration can vary widely.

You meet lasers of many varieties every day.

For example, in the supermarket checkout lane, lasers are used to understand the price of the purchased items by scanning the laser beam through the bar code. The item's price is obtained by a detector that measures the laser light reflected by the barcode.

CD and DVD players scan the discs' surface, where small pits have been burned to digitally encode the images and music that we are trying to see and hear.

Besides, lasers are also used by laser printers, which use them to transfer toner to printed pages.

Lasers have also revolutionized medicine.

The small pulsed lasers are used in surgery because they can emit energy in a very precise way and in tiny places, thus preventing unnecessary damage to nearby organs or tissues. They are also handy and minimally invasive in eye surgery, including retinal reattachment and vision correction.

The GPS

Maybe you have already heard of the Global Positioning System — or indeed its acronym, GPS — since most probably you have a navigator or cell phone.

Nevertheless, do you know that GPS would never have been

born except for the laws of quantum physics? This is because, on every GPS satellite, there is an atomic clock.

From alarm clocks to Swatch wristwatches, almost every watch counts time by recording something that occurs at a specific regular frequency.

The pendulum of a pendulum clock, for example, swings back and forth about once per second (or a frequency of one hertz), so about 60 of these equals one minute.

A modern wristwatch is based on a quartz crystal, oscillating more than 10,000 times per second. It takes many more cycles to count one minute, but the principle is the same.

We can also measure the frequency using a quantum jump in a given atom. Not the speed with which quantum jumps occur, which can be random, but the frequency carried by the photons emitted when they do.

Since this frequency is given by the difference in atomic energy levels, which is the same for each atom of the same type, it is possible to use a collection of similar atoms to maintain time.

In theory, all it takes to create an atomic clock is to know the transition frequency of your particular atom. So, you can simply sit down and figure out how many swings equals one second.

In practice, you must also make sure that your atoms are in a very stable environment so that there are no involuntary changes in their energy levels.

You can make a clock that lacks a second only once every 50 million years if you can do that!

Since the energy of each photon is very low, real-world atomic clocks need some form of amplification. Therefore, the first atomic clocks were based on the stimulated emission in the microwave range: the so-called masers.

Usually, atomic clocks use a higher power oscillator (like a quartz crystal), which is "locked" to the atomic transition frequency with an electronic feedback mechanism.

GPS itself is based on a network of satellites, each traveling in a circular orbit approximately 20,000 kilometers above the Earth's surface. Each contains an atomic clock at the same frequency. The time kept on one satellite is the time kept on another (within about one billionth of a second.) Besides, each satellite continuously transmits its position and time. The classical physics of Isaac Newton defines the position.

Meanwhile, on Earth, your GPS receiver can detect the transmission signals of at least four satellites at all times, wherever you go. The receiver then defines its position by calculating its distance from these four satellites.

This can be done because the transmitted signals travel towards the receiver at the speed of light (c) and, therefore, cover the distance (r) between the satellite and the receiver in a time $\Delta t = r/c$. The receiver measures the discrepancy between its time and that of the satellite, then calculates the distance as $r = c \times \Delta t$. This informs the receiver that it is located on a sphere of radius r centered on the satellite's known position.

When the other three satellites are considered, the receiver determines that it is located on four spheres of known rays and centers. Two of these spheres intersect in a circle, while the third sphere intersects this circle in two points. The fourth sphere determines the position unequivocally.

All of this is based on the GPS receiver, which measures time very accurately.

So, does our smartphone need its atomic clock? Fortunately, no. The GPS receiver obtains its time from the four satellites by calculating the three positions of the coordinates (x, y, z) and the time. Solving four unknown quantities with four equations is easy for even the most basic computer chip to solve four unknown quantities with four equations.

If you want to know your precise position, say less than 1 meter, you will need something more than quantum physics. You will need to add the theory of relativity. Since the satellites themselves are orbiting at such high speeds, their atomic clocks run faster than

Earth's clocks. If your receiver does not correct this, it will cause errors in the positions of a few hundred meters.

The Anti-Gravity Wheel

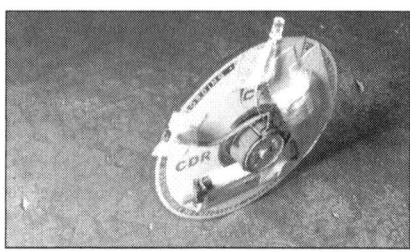

Suppose you have a 19 kg wheel attached at the end of a meter-long shaft. You could attempt to lift and hold that shaft with one hand and keep it horizontal. You could try to do it for fun, but I can guarantee that you won't be able to do it unless you are a world-class bodybuilder. If, however, you could manage to rotate the wheel on the shaft at a few thousand rpm (with the help of a drill, for example), you would then be able to raise it with one hand and hold it, even above your head, with minimum effort. The wheel would seem light as a feather.

How is this possible? This is due to the gyroscopic precession.

Instead of pulling the wheel down to the ground as one might expect, the object's weight creates torque that pushes it around, and therefore, the wheel feels light when it turns. It can be seen that the pair vector increases the angular momentum in the same direction as the couple. If there is no angular momentum at the beginning, the new momentum oscillates in the direction of the couple. If there was an angular momentum at the beginning, the direction is changed in the direction of the angular momentum, causing the precession.

The Semiconductor

For now, we have focused on atoms (and nuclei) in isolation. Look around...

Your home is full of solid objects, from this book in your hands to the ice cubes that cool in your refrigerator. Now you will know what quantum physics has to say about solids, and you will understand how quantum solids' characteristics can make our lives better.

The structure of solid matter and the atoms' arrangement in the solids derive from the number and dispositions of the electrons in each atom, i.e., from the electronic properties of the composite atoms. Although most of the solids around us are macroscopic in scale, all their physical properties are due to the electrons' movement inside them. So, solid matter and all its properties are explained by quantum physics.

Solids are divided between conductors (like the copper in household wiring) and insulators (like the paper in this book.) The difference is their ability to conduct electricity. An electrical current will flow if you connect a battery to a copper circuit. Insulators, on the other hand, do not conduct electricity. There is also an intermediate category of materials that conduct electricity only sometimes: semiconductors. Quantum physics also teaches the difference between these types of solids.

Imagine two separate atoms. Take sodium, an atom with 1 electron in its external valence shell (designated 3s), and the other 10 hidden in the shells closest to the nucleus. If well separated, each atom has its own discrete but equal set of energy levels.

If you had to bring two atoms close to each other, something strange would start happening. The wave functions of all electrons begin to overlap, and the energy levels divide into two. This happens because you now have a two-atom system, and the energy levels will be a little different depending on whether the corresponding electrons have the same spin or if they have opposite spins. Your two-atom system now has a discrete set of small pairs of energy levels.

Solids have more than two atoms. When you bring so many atoms in the immediate vicinity, the individual energy levels merge into broadband due to the wave function's overlaps and spin-spin interactions.

Returning to the sodium atom, electron dreams from well-filled internal states (1s, 2s, and 2p) will lead to well-filled energy bands. However, what about the band corresponding to single electrons in shell 3s? The valence shell for this electron can house two electrons, even if sodium has only one; i.e., the outermost shell of an isolated sodium atom is half full.

This implies that even the outermost band is only half full in the solid case. Furthermore, since the valence electrons of the atoms determined it, this band is called the valence band. A sodium atom can be excited in excited states; i.e., high energy electron shells are occupied only when the atom has been excited.

In the solid image, this means that there are also higher stretched energy bands, and these too can be occupied if the atom gets excited. Moreover, the band closest to the valence band is called the conduction band. The occupation and proximity of these two bands differentiate conductors from insulators and semiconductors.

Conductors (such as sodium) have partially filled valence bands. If an electric field is applied to a conductor, the electrons can move and occupy one of the many unoccupied energy states nearby. This movement of electrons forms an electric current, better known as electricity.

On the contrary, the valence bands of insulators and semiconductors are filled. For insulators, the energy that separates the valence and the conduction bands (called bandgap) is much more extensive than typical thermal energies.

For semiconductors, the bandgap is approximately equal to typical thermal energies. Therefore, if you add some heat to the system, the electrons near the top of the valence band can be thermally excited in the conduction band, where they can flow freely.

There have been no energetic states available to which electrons can be promoted to participate in electrical conduction.

Physicists have manipulated the unique properties of semiconductors to create many intelligent devices. For example, transistors — small electronic devices at the heart of any modern integrated circuit — rely mainly on the on-again, off-again nature of semiconductors. Silicon is among the most common semiconductors, hence the origin of the name *"Silicon Valley."*

The Solar Panel and the Light-Emitting Diode (LED)

We have already seen that some thermal energy (heat) is sufficient to promote an electron from the valence band of a semiconductor in the conduction band and to start the flow of electricity. We also know that electrons can undergo quantum leaps upwards by absorbing photons.

This remains true in the case of solids, just as the electrons in the conduction band produce electricity. The absorption of photons in a semiconductor is the basis of the photovoltaic cell, otherwise known as the solar cell.

Therefore, the quantum properties of semiconductors can be used to transform the free energy from the sun into necessary electricity (do you remember the photoelectric effect?)

The first solar cell was produced in 1954, at Bell Laboratories in the United States, by Gerald Pearson, Daryl Chapin, and Calvin Fuller.

The cell material was not made up of a single atom type.

Physicists have discovered that the electronic properties of semiconductor materials (such as silicon) can be manipulated and enhanced by injecting a small number of impurities called a dopant. If the doping atoms have more valence electrons than the base material, you will have an "n" type semiconductor. If the doping atoms have fewer electrons, you will have a "p" type semiconductor. Solar cells and transistors are based on junctions between these different types of semiconductors.

A typical solar cell, with a diameter of about 5 centimeters and a thickness of about one millimeter, can produce about 0.2 watts of energy in full sunlight.

The matrices of 50 or more cells are electrically bonded to create panels capable of producing more useful energy amounts. Continuous improvements have been made, especially to increase the fraction of sunlight converted into electricity.

The highest efficiencies obtained today are still less than 50 %; that is, half of the sunlight still manages to heat the solar cells before getting lost in the environment.

Can solids emit light when electrons drop from higher to lower energy levels like atoms? The answer is affirmative, and this is the basis for the light-emitting diode (LED.)

Today, LEDs are omnipresent in the displays of our stereos, watches, and appliances. LED lighting replaced the incandescent bulb, an innovation so precious that it won the 2014 Nobel Prize in physics.

Superconductivity

As the name already says, the only thing better than a semiconductor is a superconductor. Superconductors take their name because they can carry supercurrents, electric currents that flow without electrical resistance.

We have already talked about the current movement of electric charges in a conductive material.

We also talked about resistance. An electrical "friction" leads to unwanted loss of current by conversion to heat.

When connecting a battery to a simple (non-superconducting) circuit, the circuit's resistance defines the amount of current that can flow. Suppose you increase the resistance, the current decreases. The circuit's resistance draws energy from the electric current and converts it into a less useful heat form.

When you cool a superconductor, there is a low temperature below which the resistance drops immediately to zero. This is an excellent property since energy lost as heat in ordinary circuits can be preserved in their superconducting counterparts.

Quantum physics provides us with an understanding of what is going on.

Electrical resistance appears in most natural materials because every flowing electron encounters something during its journey. For example, this could be a positive ion in the underlying material, an impurity atom, or an imperfection in the crystal structure. Whenever an electron collides with one of these things, it loses some of its energy as heat.

Classical physics would say that your collisions will remain no matter how cool you are. Therefore, it cannot explain superconductivity. Instead, it is not necessary to consider the individual properties of the particles of an electron but the collective wavy properties of many electrons in the solid. In the superconducting state, all electrons in the material form a single coherent wave function. Once in this state, the crystal's impurities and imperfections become negligible impediments, and the current can flow without resistance.

Although zero resistance conductors appear to offer many advantages, they have some drawbacks. The first is to keep the circuit fresh and pleasant, usually below 200 ° C. It is so cold that special cryogenic liquids are needed.

Fortunately, some "high temperature" superconductors have been discovered, although we must stress that "high" is a relative term. These still need to be cooled below about -100° C to enter the superconducting state.

They are fragile materials. Therefore, they are not the best choice for electrical wiring. Therefore, superconducting power lines will take some time to transport electricity to our homes and reduce our electricity bills.

Superconductors have another attractive property: they expel external magnetic fields when they are superconducting. This function is called the Meissner effect. Many technologies use this property.

For example, magnetic levitation. If a magnet is placed just above a superconducting material, the force applied by the superconductor to expel the magnetic field from its interior will allow the magnet to float in the space above. It also helped in the development of a superfast and efficient train. The "maglev" train was designed to float on some form of superconducting "guide" using magnets rather than wheels, axles, and bearings.

Today, we are looking for new and improved superconducting materials, especially those that can operate at much more comfortable temperatures.

Quantum Physics as Seen in Everyday Objects

At a point, we must feel irritated and slightly confused about how so many concepts mentioned here can be applied to our daily lives or used by the different instruments around us. Quantum Physics is one of the highlights of human intellectualism, and its knowledge has helped shape our civilization. Despite this relevance, most people still feel the subject of this field is quite abstruse and cannot be easily grasped by the ordinary mind. In the mind of the public, the concept of quantum physics is seen as a hard concept that is only understood by minds like Einstein and Hawking and another superhuman brain.

The concept of quantum physics is an understanding of the universe, and the universe is all around us, and its operation is based on quantum rules. Even though we are so used to the laws of classical physics, and this relates to the universe at a macroscopic level, the understanding of quantum physics still affects various familiar operations. You'll find this list contains various tools and equipment that apply to the quantum principle without us realizing it.

Toasters

We are all familiar with the red glow produced by the heating element as we toast our bread. Funny enough, it was the observation of this red light that led physicists to ask questions, questions that birthed the quantum concept. Physicists wanted to know why hot objects shone that particular color of red, a very tough question, and quantum physics came to provide light to it.

Max Planck answered this issue in his theory, where he said that the light been transmitted must be discharged in discrete pieces of vitality, actual products of short, consistent occasions the recurrence of the light. For high-frequency light, the energy quantum is greater than the share of heat energy, which is assigned to that frequency, and this makes it impossible for light to be emitted at such frequency. This prevents the emission of high-frequency light. We could say that the toaster could be a central place where the idea of quantum physics first originated.

Fluorescent Lights

The traditional incandescent light bulb was able to emit light by heating a piece of wire adequately until it gets hot and emit a bright white glow; this is similar to the phenomenon of the toaster. You enjoy groundbreaking quantum physics work whenever you flick on a fluorescent bulb or one of the more recent twisty CFL bulbs; that's quantum physics at work.

In the early 19th century, physicists discovered that all elements found in the periodic table have a unique spectrum. When we heat a vapor of atoms, they will eventually emit light at a small number of discrete wavelengths, and each of the different items will have a different pattern. The spectral lines were used to classify the composition of new material, and unknown elements such as helium were first discovered through this process.

This is how a fluorescent bulb works: whether the bulb is CFL or long tube, inside the bulb is a tiny bit of mercury vapor that is excited into plasma. Mercury easily emits light at frequencies that fall at the visible spectrum, and eyes will perceive this as white light.

Afterword

Thank you for making it through to the end of this book. Quantum physics is an endlessly fascinating subject and one that we think everyone should learn about in some capacity. After all, as we discover more and more about quantum physics, more of the future is open to us — things like teleportation, supercomputing, and ultra-fast space travel all feel like they're just around the corner if we can finally unravel how quantum mechanics works. Of course, we're not all scientists, and we can't all conduct experiments to figure out the next big discovery, but we can all do our part just by learning a little bit about how the universe around us works.

There are so many things going on in the universe around us that we can't explain. However, there's beauty in this un-knowing, and there's an indelible fascination in the discovery of the world around us. How boring would it be to know already exactly how everything around us worked? If we knew that, we would be able to predict everything, from whether there was alien life in the far reaches of the universe to when our planet and solar system would die. While some might argue that it might be better to know, the universe's vastness allows for infinite new things to discover, crazy new laws and theorems to figure out, and new phenomena just over the event horizon. This information has assisted in the understanding of how stars are born and what force and matter do when they interact with each other on a particle level and also in larger masses.

The next step is to incorporate the knowledge you've gained here into your everyday life. Whether you use your new knowledge simply to brag, to help others, to do your research, to deepen your understanding of our world, or to become the next Albert Einstein, we're sure it will enrich your life in new and exciting ways. Please never stop reading, and more importantly, never stop learning.

Made in the USA
Monee, IL
16 August 2022